Hidden Semi-Markov Models

Hidden Semi-Markov Models

Theory, Algorithms and Applications

Shun-Zheng Yu

AMSTERDAM • BOSTON • HEIDELBERG • LONDON • NEW YORK
OXFORD • PARIS • SAN DIEGO • SAN FRANCISCO • SINGAPORE • SYDNEY • TOKYO

Elsevier
Radarweg 29, PO Box 211, 1000 AE Amsterdam, Netherlands
The Boulevard, Langford Lane, Kidlington, Oxford OX5 1GB, UK
225 Wyman Street, Waltham, MA 02451, USA

ISBN: 978-0-12-802767-7

British Library Cataloguing-in-Publication Data
A catalogue record for this book is available from the British Library

Library of Congress Cataloging-in-Publication Data
A catalog record for this book is available from the Library of Congress

For information on all Elsevier publications
visit our website at http://store.elsevier.com/

CONTENTS

Preface vii
Acknowledgments xi

Chapter 1 Introduction **1**
1.1 Markov Renewal Process and Semi-Markov Process 1
1.2 Hidden Markov Models 9
1.3 Dynamic Bayesian Networks 13
1.4 Conditional Random Fields 19
1.5 Hidden Semi-Markov Models 22
1.6 History of Hidden Semi-Markov Models 24

Chapter 2 General Hidden Semi-Markov Model **27**
2.1 A General Definition of HSMM 27
2.2 Forward–Backward Algorithm for HSMM 32
2.3 Matrix Expression of the Forward–Backward Algorithm 41
2.4 Forward-Only Algorithm for HSMM 46
2.5 Viterbi Algorithm for HSMM 51
2.6 Constrained-Path Algorithm for HSMM 54

Chapter 3 Parameter Estimation of General HSMM **59**
3.1 EM Algorithm and Maximum-Likelihood Estimation 59
3.2 Re-estimation Algorithms of Model Parameters 67
3.3 Order Estimation of HSMM 72
3.4 Online Update of Model Parameters 74

Chapter 4 Implementation of HSMM Algorithms **85**
4.1 Heuristic Scaling 85
4.2 Posterior Notation 88
4.3 Logarithmic Form 91
4.4 Practical Issues in Implementation 93

Chapter 5 Conventional HSMMs **103**
5.1 Explicit Duration HSMM 103
5.2 Variable Transition HSMM 108

5.3 Variable-Transition and Explicit-Duration Combined HSMM 113
5.4 Residual Time HSMM 114

Chapter 6 Various Duration Distributions 121
6.1 Exponential Family Distribution of Duration 121
6.2 Discrete Coxian Distribution of Duration 123
6.3 Duration Distributions for Viterbi HSMM Algorithms 127

Chapter 7 Various Observation Distributions 129
7.1 Typical Parametric Distributions of Observations 129
7.2 A Mixture of Distributions of Observations 131
7.3 Multispace Probability Distributions 135
7.4 Segmental Model 137
7.5 Event Sequence Model 140

Chapter 8 Variants of HSMMs 143
8.1 Switching HSMM 143
8.2 Adaptive Factor HSMM 145
8.3 Context-Dependent HSMM 146
8.4 Multichannel HSMM 153
8.5 Signal Model of HSMM 155
8.6 Infinite HSMM and HDP-HSMM 157
8.7 HSMM Versus HMM 159

Chapter 9 Applications of HSMMs 163
9.1 Speech Synthesis 165
9.2 Human Activity Recognition 172
9.3 Network Traffic Characterization and Anomaly Detection 175
9.4 fMRI/EEG/ECG Signal Analysis 177

References 179

PREFACE

A hidden semi-Markov model (HSMM) is a statistical model. In this model, an observation sequence is assumed to be governed by an underlying semi-Markov process with unobserved (hidden) states. Each hidden state has a generally distributed duration, which is associated with the number of observations produced while in the state, and a probability distribution over the possible observations.

Based on this model, the model parameters can be estimated/updated, the predicted, filtered, and smoothed probabilities of partial observation sequence can be determined, goodness of the observation sequence fitting to the model can be evaluated, and the best state sequence of the underlying semi-Markov process can be found.

Due to those capabilities of the HSMM, it becomes one of the most important models in the area of artificial intelligence/machine learning. Since the HSMM was initially introduced in 1980 for machine recognition of speech, it has been applied in more than forty scientific and engineering areas with thousands of published papers, such as speech recognition/synthesis, human activity recognition/prediction, network traffic characterization/anomaly detection, fMRI/EEG/ECG signal analysis, equipment prognosis/diagnosis, etc.

Since the first HSMM was introduced in 1980, three other basic HSMMs and several variants of them have been proposed in the literature, with various definitions of duration distributions and observation distributions. Those models have different expressions, algorithms, computational complexities, and applicable areas, without explicitly interchangeable forms. A unified definition, in-depth treatment and foundational approach of the HSMMs are in strong demand to explore the general issues and theories behind them.

However, in contrast to a large number of published papers that are related to HSMMs, there are only a few review articles/chapters on HSMMs, and none of them aims at filling the demand. Besides, all existing review articles/chapters were published several years ago. New developments and emerging topics that have surfaced in this field need to be summarized.

Therefore, this book is intended to include the models, theory, methods, applications, and the latest information on development in this field. In summary, this book will provide:

- a unified definition, in-depth treatment and foundational approach of the HSMMs;
- a survey on the latest development and emerging topics in this field;
- examples helpful for the general reader, teachers and students in computer science, and engineering to understand the topics;
- a brief description of applications in various areas;
- an extensive list of references to the HSMMs.

For these purposes, this book presents nine chapters in three parts. In the first part, this book defines a unified model of HSMMs, and discusses the issues related to the general HSMM, which include:

1. the forward–backward algorithms that are the fundamental algorithms of HSMM, for evaluating the joint probabilities of partial observation sequence;
2. computation of the predicted/filtered/smoothed probabilities, expectations, and the likelihood function of observations, which are necessary for inference in HSMM;
3. the maximum a posteriori estimation of states and the estimation of best state sequence by Viterbi HSMM algorithm;
4. the maximum-likelihood estimation, training and online update of model parameters; proof of the re-estimation algorithms by the EM algorithm;
5. practical issues in the implementation of the forward–backward algorithms.

By introducing certain assumptions and some constraints on the state transitions, the general HSMM becomes the conventional HSMMs, including explicit duration HMM, variable transition HMM, and residual time HMM. Those conventional models have different capability in modeling applications, with different computational complexity and memory requirement involved in the forward–backward algorithms and the model estimation.

In the second part, this book discusses the state duration distributions and the observation distributions, which can be nonparametric or parametric depending on the specific preference of the applications.

Among the parametric distributions, the most popular ones are the exponential family distributions, such as Poisson, exponential, Gaussian, and gamma. A mixture of Gaussian distributions is also widely used to express complex distributions.

Other than the exponential family and the mixed distributions, the Coxian distribution of state duration can represent any discrete probability density function, and the underlying series–parallel network also reveals the structure of different HSMMs.

A multispace probability distribution is applied to express a composition of different dimensional observation spaces, or a mixture of continuous and discrete observations. A segmental model of observation sequence is used to describe parametric trajectories that change over time. An event sequence model is used to model and handle an observation sequence with missed observations.

In the third part, this book discusses variants and applications of HSMMs. Among the variants of HSMMs, a switching HSMM allows the model parameters to be changed in different time periods. An adaptive factor HSMM allows the model parameters to be a function of time. A context-dependent HSMM lets the model parameters be determined by a given series of contextual factors. A multichannel HSMM describes multiple interacting processes. A signal model of HSMM uses an equivalent form to express an HSMM.

There usually exists a class of HSMMs that are specified for the applications in an area. For example, in the area of speech synthesis, speech features (observations to be obtained), instead of the model parameters, are to be determined. In the area of human activity recognition, unobserved activity (hidden state) is to be estimated. In the area of network traffic characterization/anomaly detection, performance/health of the entire network is to be evaluated. In the area of fMRI/EEG/ECG signal analysis, neural activation is to be detected.

ACKNOWLEDGMENTS

I would like to thank Dr Yi XIE, Bai-Chao LI, and Jian-Zheng LUO, who collected a lot of papers that are related to HSMMs, and sorted them based on the relevancy to the applicable theories, algorithms, and applications. Their work is instrumental for me to complete the book in time. I also want to express my gratitude toward the reviewers who carefully read the draft and provide me with many valuable comments and suggestions. Dr Yi XIE, Bai-Chao LI, Wei-Tao WU, Qin-Liang LIN, Xiao-Fan CHEN, Yan LIU, Jian KUANG, and Guang-Rui WU proofread the chapters. Without their tremendous effort and help, it would have been extremely difficult for me to finish this book as it is now.

CHAPTER 1

Introduction

This chapter reviews some topics that are closely related to hidden semi-Markov models, and introduces their concepts and brief history.

1.1 MARKOV RENEWAL PROCESS AND SEMI-MARKOV PROCESS

In this chapter, we briefly review the Markov renewal process and semi-Markov process, as well as generalized semi-Markov process and discrete-time semi-Markov process.

1.1.1 Markov Renewal Process

A renewal process is a generalization of a Poisson process that allows arbitrary holding times. Its applications include such as planning for replacing worn-out machinery in a factory. A Markov renewal process is a generalization of a renewal process that the sequence of holding times is not independent and identically distributed. Their distributions depend on the states in a Markov chain. The Markov renewal processes were studied by Pyke (1961a, 1961b) in 1960s. They are applied in M/G/1 queuing systems, machine repair problem, etc. (Cinlar, 1975).

Denote $\mathbf{S}$ as a state space and $X_n \in \mathbf{S}$, $n = 0, 1, 2, \ldots,$ as states in the Markov chain. Let (X_n, T_n) be a sequence of random variables, where T_n are the jump times of the states. The inter-arrival times of the states are $\tau_n = T_n - T_{n-1}$. If

$$\begin{aligned} P[\tau_{n+1} \le \tau, X_{n+1} = j | (X_0, T_0), (X_1, T_1), \ldots, (X_n = i, T_n)] \\ = P[\tau_{n+1} \le \tau, X_{n+1} = j | X_n = i], \end{aligned}$$

for any $n \ge 0$, $\tau \ge 0$, $i, j \in \mathbf{S}$, then the sequence (X_n, T_n) is called a Markov renewal process. In other words, in a Markov renewal process, the next state $X_{n+1} = j$ and the inter-arrival time τ_{n+1} to the next state is dependent on the current state $X_n = i$ and independent of the historical states $X_0, X_1, \ldots, X_{n-1}$ and the jump times $T_1, \ldots, T_n$. Define the state transition probabilities by

$$H_{ij}(\tau) = P[\tau_{n+1} \leq \tau, X_{n+1} = j | X_n = i]$$

with $H_{ij}(0) = 0$, that is, at any time epoch multiple transitions are not allowed. Let

$$N(t) = \max\{n : T_n \leq t\}$$

count the number of renewals in the interval $[0, t]$. Then $N(t) < \infty$ for any given $t \geq 0$. $(N(t))_{t \geq 0}$ is called a Markov renewal counting process.

1.1.2 Semi-Markov Process

A semi-Markov process is equivalent to a Markov renewal process in many aspects, except that a state is defined for every given time in the semi-Markov process, not just at the jump times. Therefore, the semi-Markov process is an actual stochastic process that evolves over time. Semi-Markov processes were introduced by Levy (1954) and Smith (1955) in 1950s and are applied in queuing theory and reliability theory.

For an actual stochastic process that evolves over time, a state must be defined for every given time. Therefore, the state S_t at time t is defined by $S_t = X_n$ for $t \in [T_n, T_{n+1})$. The process $(S_t)_{t \geq 0}$ is thus called a semi-Markov process. In this process, the times $0 = T_0 < T_1 < \cdots < T_n < \cdots$ are the jump times of $(S_t)_{t \geq 0}$, and $\tau_n = T_n - T_{n-1}$ are the sojourn times in the states. Every transition from a state to the next state is instantaneously made at the jump times.

For a time-homogeneous semi-Markov process, the transition density functions are

$$h_{ij}(\tau) d\tau \equiv P[\tau \leq \tau_{n+1} < \tau + d\tau, X_{n+1} = j | X_n = i],$$

where $h_{ij}(\tau)$ is independent of the jumping time T_n. It is the probability density function that after having entered state i at time zero the process transits to state j in between time τ and $\tau + d\tau$. They must satisfy

$$\sum_{j \in \mathbf{S}} \int_0^{\infty} h_{ij}(\tau) d\tau = 1,$$

for all $i \in \mathbf{S}$. That is, state i must transit to another state in the time $[0, \infty)$.

If the number of jumps in the time interval $[0, T]$ is $N(T) = n$, then the sample path $(s_t, t \in [0, T])$ is equivalent to the sample path $\left(x_0, \tau_1, x_1, \ldots, \tau_n, x_n, T - \sum_{k=1}^{n} \tau_k\right)$ with probability 1. Then the joint distribution of the process $(s_t)_{0 \le t \le T}$ is

$$P\left[x_0, \tau_1' \le \tau_1, x_1, \ldots, \tau_n' \le \tau_n, x_n, T - \sum_{k=1}^{n} \tau_k' | N(T) = n\right]$$

$$= P[x_0] \cdot \int_0^{\tau_1} \cdots \int_0^{\tau_n} \left(1 - W_{x_n}\left(T - \sum_{k=1}^{n} \tau_k'\right)\right) \cdot \prod_{k=1}^{n} h_{x_{k-1}x_k}(\tau_k') \cdot d\tau_k',$$

where $W_i(\tau) = \int_0^{\tau} \sum_{j \in \mathbf{S}} h_{ij}(\tau') d\tau'$ is the probability that the process stays in state i for at most time τ before transiting to another state, and $1 - W_i(\tau)$ is the probability that the process will not make transition from state i to any other state within time τ. The likelihood function corresponding to the sample path $(x_0, \tau_1, x_1, \ldots, \tau_n, x_n, T - \sum_{k=1}^{n} \tau_k)$ is thus

$$L\left[x_0, \tau_1, x_1, \ldots, \tau_n, x_n, T - \sum_{k=1}^{n} \tau_k\right] = P[x_0] \cdot \left(1 - W_{x_n}\left(T - \sum_{k=1}^{n} \tau_k\right)\right) \cdot \prod_{k=1}^{n} h_{x_{k-1}x_k}(\tau_k).$$

Suppose the current time is t. The time that has been passed since last jump is defined by $R_t = t - T_{N(t)}$. Then the process (S_t, R_t) is a continuous time homogeneous Markov process.

The semi-Markov process can be generated by different types of random mechanisms (Nunn and Desiderio, 1977), for instances:

1. Usually, it is thought as such a stochastic process that after having entered state i, it randomly determines the successor state j based on the state transition probabilities a_{ij}, and then randomly determines the amount of time τ staying in state i before going to state j based on the holding time density function $f_{ij}(\tau)$, where

$a_{ij} \equiv P[X_{n+1} = j | X_n = i] = \int_0^\infty h_{ij}(\tau)d\tau$ is the transition probability from state i to state j, s.t. $\sum_{j \in \mathbf{S}} a_{ij} = 1$, and

$$f_{ij}(\tau)d\tau \equiv P[\tau \leq \tau_{n+1} < \tau + d\tau | X_n = i, X_{n+1} = j] = h_{ij}(\tau)d\tau / a_{ij}$$

is the probability that the transition to the next state will occur in the time between τ and $\tau + d\tau$ given that the current state is i and the next state is j. In this model,

$$h_{ij}(\tau) = a_{ij} f_{ij}(\tau).$$

2. The semi-Markov process can be thought as a stochastic process that after having entered state i, it randomly determines the waiting time τ for transition out of state i based on the waiting time density function $w_i(\tau)$, and then randomly determines the successor state j based on the state transition probabilities $a_{(i,\tau)j}$, where $w_i(\tau)$ is the density function of the waiting time for transition out of state i defined by

$$w_i(\tau)d\tau = P[\tau \leq \tau_{n+1} < \tau + d\tau | X_n = i] = \sum_{j \in \mathbf{S}} h_{ij}(\tau)d\tau,$$

and

$$a_{(i,\tau)j} \equiv P[X_{n+1} = j | X_n = i, \tau_{n+1} = \tau]$$

is the probability that the system will make the next transition to state j, given time τ and current state i. In this model,

$$h_{ij}(\tau) = w_i(\tau) a_{(i,\tau)j}.$$

3. The semi-Markov process can also be thought as such a process that after having entered state i, it randomly draws the pair (k, d_{ik}) for all $k \in \mathbf{S}$, based on $f_{ik}(\tau)$, and then determines the successor state and length of time in state i from the smallest draw. That is, if $d_{ij} = \min_{k \in \mathbf{S}}\{d_{ik}\}$, then the next transition is to state j and the length of time the process holds in state i before going to state j is d_{ij}. In this model,

$$h_{ij}(\tau) = f_{ij}(\tau) \prod_{k \neq j} (1 - F_{ik}(\tau)),$$

where $F_{ik}(\tau) = \int_0^\tau f_{ik}(\tau')d\tau'$, and $\prod_{k \neq j}(1 - F_{ik}(\tau))$ is the probability that the process will not transit to another state except j by time τ. This type of semi-Markov process is applied to such as reliability analysis (Veeramany and Pandey, 2011). An example of this type of semi-Markov process is as follows.

Example 1.1

Suppose a multiple-queue system contains L queues, each with known inter-arrival time distribution and departure time distribution. Let q_l be the length of the lth queue and define the state at time t by $S_t = (q_1, \ldots, q_L)$. Then every external arrival to a queue will increment the queue length and every departure from a queue will decrement the queue length if it is greater than zero. Therefore, each arrival/departure will result in a transition of the system to a corresponding state. Denote an external arrival to queue l by $e_l^+ = (0, \ldots, 1, \ldots, 0)$ and a departure from queue l with $q_l > 0$ by $e_l^- = (0, \ldots, -1, \ldots, 0)$. Then the next state is $(q_1, \ldots, q_L) + e_l^+$ or $(q_1, \ldots, q_L) + e_l^-$ for $l = 1, \ldots, L$. The time to the next state transition is determined by which e_l^+ or e_l^-, for $l = 1, \ldots, L$, occurs first, based on their inter-arrival/departure time distributions.

1.1.3 Generalized Semi-Markov Process

A generalized semi-Markov process extends a semi-Markov process by letting an event trigger a state transition and the next state be determined by the current state and the event that just occurred. It is applied to discrete event systems (Glynn, 1989).

In Example 1.1, if the multiple queues form an open queuing network, such as in a packet switching network, a data packet sent out of a switch will go out of the network or randomly select one of its neighbor switches to enter. Every switch is assumed to have an input queue, as shown in Figure 1.1. Therefore, a state transition depends on the current state as well as the next arrival/departure event. The semi-Markov process is thus extended to so-called *Generalized Semi-Markov Process*.

Suppose that the current state is $S_t = S_{T_n} = X_n$, and the next event is E_{n+1} that will cause the transition into next state X_{n+1} at time epoch T_{n+1}. Denote $\mathbf{E}(i)$ as the set of events that can cause the process being out of state i. Then

$$\begin{aligned} h_{ij}(\tau)d\tau &\equiv P[\tau \le \tau_{n+1} < \tau + d\tau, X_{n+1} = j | X_n = i] \\ &= \sum_{e \in \mathbf{E}(i)} P[\tau_{n+1} \in d\tau, X_{n+1} = j, E_{n+1} = e | X_n = i] \\ &= \sum_{e \in \mathbf{E}(i)} P[\tau_{n+1} \in d\tau, E_{n+1} = e | X_n = i] \cdot P[X_{n+1} = j | X_n = i, E_{n+1} = e], \end{aligned}$$

where $\tau_{n+1} \in d\tau$ represents $\tau \le \tau_{n+1} < \tau + d\tau$. Therefore, the transition probability is extended to $P[X_{n+1} = j | X_n = i, E_{n+1} = e]$, that is, the next

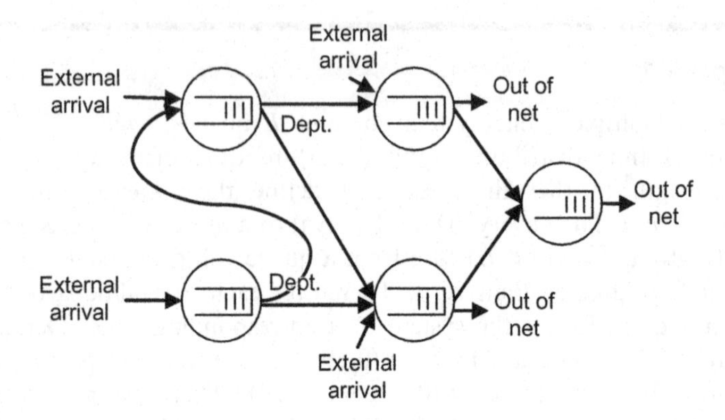

Figure 1.1 Open queuing network.
There are five switches in the open network. Each switch has a queue of first-in-first-out. When a switch receives a packet, regardless it is an external arrival or a departure from another switch, the packet will be input into its queue. The packet in the "head" of the queue will be sent out of the network or to one of other switches. The vector of the lengths of those five queues is treated as the state. Therefore, each external arrival or a departure from a queue will change the state.

transition depends on both the current state and the next event. The next transition epoch T_{n+1} is determined by the event $E_{n+1} \in \mathbf{E}(i)$ that occurs first among all events of $\mathbf{E}(i)$. That is,

$$P[\tau_{n+1} \in d\tau, E_{n+1} = e | X_n = i] = P[\tau_{n+1} \in d\tau | X_n = i, E_{n+1} = e] \cdot \prod_{\substack{e' \in \mathbf{E}(i) \\ e' \neq e}} P[T_{e',next} - T_n > \tau | X_n = i, E_{n+1} = e'],$$

where $T_{e',\text{next}}$ denotes the time epoch that event e' will occur. In considering that each event has an inter-event time distribution, the inter-event time for event e has passed $c_{e,n}$ at time epoch T_n since event e lastly occurred. Suppose $T_{e,\text{last}}$ and $T_{e,\text{next}}$, for $T_{e,\text{last}} \leq T_n < T_{e,\text{next}}$, are the epochs that event e lastly occurred and will appear at. The inter-event time for event e is thus $y_{e,n} = T_{e,\text{next}} - T_{e,\text{last}}$. Then,

$$P[\tau_{n+1} \in d\tau | X_n = i, E_{n+1} = e] = \frac{P[c_{e,n} + \tau \leq y_{e,n} < c_{e,n} + \tau + d\tau | X_n = i, E_{n+1} = e]}{P[y_{e,n} > c_{e,n} | X_n = i, E_{n+1} = e]}$$

and

$$P[T_{e',next} - T_n > \tau | X_n = i, E_{n+1} = e'] = \frac{P[c_{e',n} + \tau < y_{e',n} | X_n = i, E_{n+1} = e']}{P[y_{e',n} > c_{e',n} | X_n = i, E_{n+1} = e']}.$$

Example 1.1 (continued)

In the open queuing network system, a departure from queue l can occur only if $q_l > 0$. That is, not all events can occur while the system is in a given state. Then for state $X_n = x_n$, the event set that can cause the system out of state x_n is

$$\mathbf{E}(x_n) = \{e_l^+ : l = 1, \ldots, L\} \cup \{e_l^- : x_n(l) > 0, l = 1, \ldots, L\},$$

where $x_n(l) = q_l$ is the lth element of x_n. A data packet departed from queue l will randomly select queue k to enter with probability p_{lk}, and go out of the system with probability $1 - \sum_{k=1}^{L} p_{lk}$. Therefore, for the given event e_l^-, the former results in that the system transits from state $X_n = x_n$ to state $X_{n+1} = x_n + e_l^- + e_k^+$ with probability p_{lk}, and the latter to state $X_{n+1} = x_n + e_l^-$ with probability $1 - \sum_{k=1}^{L} p_{lk}$. For an external arrival event e_l^+, the probability from state $X_n = x_n$ to state $X_{n+1} = x_n + e_l^+$ is 1.

1.1.4 Discrete-Time Semi-Markov Process

If t is discrete, that is, $t = 0, 1, 2, \ldots$, then $(S_t)_{t \geq 0}$ is called a discrete-time semi-Markov process. In this case, if t_1 is the starting time of state $j \in \mathbf{S}$ and t_2 its ending time, then $S_{t_1} = j$, $S_{t_1+1} = j$, $\ldots$, $S_{t_2} = j$ with $S_{t_1-1} \neq j$ and $S_{t_2+1} \neq j$. The sojourn time of $(S_t)_{t \geq 0}$ in state j is an integer $\tau = t_2 - t_1 + 1 \geq 1$. In this case, the transition mass functions are

$$h_{ij}(\tau) \equiv P[\tau_{n+1} = \tau, X_{n+1} = j | X_n = i].$$

Define

$$h_{ij}^{(m)}(\tau) \equiv P[\tau_{n+m} = \tau, X_{n+m} = j | X_n = i]$$

as the probability that, starting from state i at time T_n, the semi-Markov process will do the mth jump at time $T_n + \tau$ to state j. It must satisfy $\tau \geq m$. In other words, within a finite time τ, the semi-Markov process can make at most τ jumps, that is,

$$\sum_{m=1}^{\infty} h_{ij}^{(m)}(\tau) = \sum_{m=1}^{\tau} h_{ij}^{(m)}(\tau).$$

It is due to this fact that the discrete-time semi-Markov process is different from the continuous-time semi-Markov process (Barbu and Limnios, 2008), which can make infinite jumps within time τ.

Similarly define the state transition probabilities of a discrete-time semi-Markov process by

$$g_{ij}(\tau) = P[S_{t+\tau} = j | S_t = i]$$

and the cumulative distribution function of waiting time in state i by

$$W_i(\tau) = \sum_{j \in \mathbf{S}} \sum_{\tau'=1}^{\tau} h_{ij}(\tau'),$$

where $1 - W_i(\tau)$ is the probability that the sojourn time of state i is at least τ. Then the Markov renewal equation for the transition probabilities of discrete-time semi-Markov process is

$$g_{ij}(\tau) = (1 - W_i(\tau)) \cdot I(i = j) + \sum_{k \in \mathbf{S}} \sum_{l=1}^{\tau} h_{ik}(l) g_{kj}(\tau - l),$$

where $I(i = j)$ is an indicator function which equals 1 if $i = j$; otherwise 0. This recursive formula can be used to compute the transition functions $g_{ij}(\tau)$ of the discrete-time semi-Markov process.

Example 1.2

There are two libraries. Readers can borrow books from and return to any of them, as shown in Figure 1.2. The statistics of the libraries shows that the transition mass functions are

$$h_{1,1}(\tau) = 0.24 \times 0.7^{\tau-1}, \quad h_{1,2}(\tau) = 0.04 \times 0.8^{\tau-1}$$
$$h_{2,1}(\tau) = 0.12 \times 0.6^{\tau-1}, \quad h_{2,2}(\tau) = 0.07 \times 0.9^{\tau-1}.$$

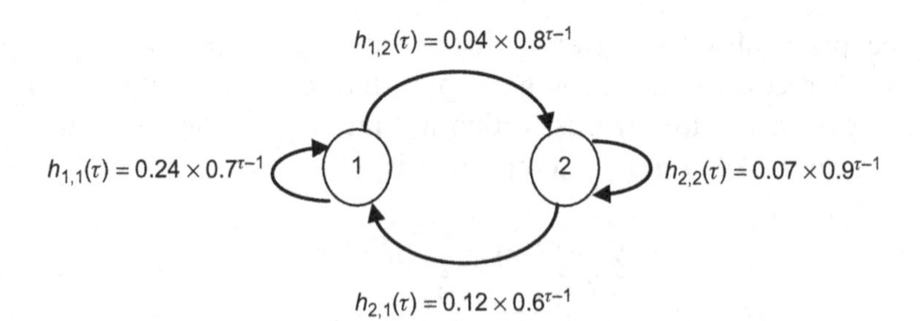

Figure 1.2 The semi-Markov process.
There are two states. Each state can transit to itself or the other one with transition mass functions. For example, state 1 can transit to state 1 with transition mass function $h_{1,1}(\tau)$ and state 2 with $h_{1,2}(\tau)$.

Then we can get the probability that a book is borrowed from library i and returned to library j by $a_{ij} = \sum_{\tau=1}^{\infty} h_{ij}(\tau)$. It yields that

$$a_{1,1} = 0.24/0.3 = 0.8 \quad a_{1,2} = 0.04/0.2 = 0.2$$
$$a_{2,1} = 0.12/0.4 = 0.3 \quad a_{2,2} = 0.07/0.1 = 0.7.$$

The holding time probabilities after a book is borrowed from library i and decided to return to library j are $f_{ij}(\tau) = h_{ij}(\tau)/a_{ij}$, that is,

$$f_{1,1}(\tau) = 0.3 \times 0.7^{\tau-1} \quad f_{1,2}(\tau) = 0.2 \times 0.8^{\tau-1}$$
$$f_{2,1}(\tau) = 0.4 \times 0.6^{\tau-1} \quad f_{2,2}(\tau) = 0.1 \times 0.9^{\tau-1}.$$

The mean holding time that a book is borrowed from library i and returned to library j is $\bar{d}_{ij} = \sum_{\tau=1}^{\infty} f_{ij}(\tau)\tau$. That is,

$$\bar{d}_{1,1} = 1/0.3 \approx 3.33 \quad \bar{d}_{1,2} = 1/0.2 = 5$$
$$\bar{d}_{2,1} = 1/0.4 = 2.5 \quad \bar{d}_{2,2} = 1/0.1 = 10.$$

These show that a book borrowed from library 2 and returned to library 2 often has longer holding period. Thus the mean holding times that a book is borrowed from different libraries are

$$\bar{d}_1 = a_{1,1}\bar{d}_{1,1} + a_{1,2}\bar{d}_{1,2} \approx 3.67$$
$$\bar{d}_2 = a_{2,1}\bar{d}_{2,1} + a_{2,2}\bar{d}_{2,2} = 7.75.$$

These show that people hold a book borrowed from library 2 for a longer time.

1.2 HIDDEN MARKOV MODELS

In this chapter, we briefly review the hidden Markov model (HMM). An HMM is defined as a doubly stochastic process. The underlying stochastic process is a discrete-time finite-state homogeneous Markov chain. The state sequence is not observable and so is called hidden. It influences another stochastic process that produces a sequence of observations. The HMM was first proposed by Baum and Petrie (1966) in the late 1960s. An excellent tutorial of HMMs can be found in Rabiner (1989), a theoretic overview of HMMs can be found in Ephraim and Merhav (2002) and a discussion on learning and inference of HMMs in understanding of Bayesian networks (BNs) is presented in Ghahramani (2001).

Assume a homogeneous (i.e., time-invariant) discrete-time Markov chain with a set of (hidden) states $\mathbf{S} = \{1, \ldots, M\}$. The state sequence is denoted by $S_{1:T} \equiv (S_1, \ldots, S_T)$, where $S_t \in \mathbf{S}$ is the state at time t. A realization of $S_{1:T}$ is denoted as $s_{1:T}$. Define $a_{ij} \equiv P[S_t = j | S_{t-1} = i]$ as the transition probability from state i to state j, and $\pi_j \equiv P[S_0 = j]$ the initial distribution of the state.

Denote the observation sequence by $O_{1:T} \equiv (O_1, \ldots, O_T)$, where $O_t \in \mathbf{V}$ is the observation at time t and $\mathbf{V} = \{v_1, v_2, \ldots, v_K\}$ is the set of observable values. The emission probability of observing v_k while transiting from state i to state j is denoted by $b_{ij}(v_k) \equiv P[v_k | S_{t-1} = i, S_t = j]$. It is often assumed in the literature that the observation is independent of the previous state and hence

$$b_{ij}(v_k) = b_j(v_k) \equiv P[v_k | S_t = j].$$

Therefore, the set of model parameters of an HMM is $\lambda \equiv \{a_{ij}, b_{ij}(v_k), \pi_j : i, j \in \mathbf{S}, v_k \in \mathbf{V}\}$. The standard HMM is explained in Figure 1.3.

Given the set of model parameters λ and an instance of the observation $O_{1:T} = o_{1:T}$, the probability that this observed sequence is generated by the model is $P[o_{1:T} | \lambda]$. A computation method for $P[o_{1:T} | \lambda]$ using the sum–product expression is

$$\begin{aligned} P[o_{1:T} | \lambda] &= \sum_{j \in S} P[S_t = j, o_{1:T} | \lambda] \\ &= \sum_{j \in S} P[S_t = j, o_{1:t} | \lambda] P[o_{t+1:T} | S_t = j, \lambda] \\ &= \sum_{j \in S} \alpha_t(j) \beta_t(j), \end{aligned}$$

where

$$\alpha_t(j) \equiv P[S_t = j, o_{1:t} | \lambda]$$

is the *forward variable* defined as the joint probability of $S_t = j$ and the partial observed sequence $o_{1:t}$, and

$$\beta_t(j) \equiv P[o_{t+1:T} | S_t = j, \lambda]$$

the *backward variable* defined as the probability of future observations given the current state. In the derivation of the sum–product expression, the Markov property that the future observations are dependent

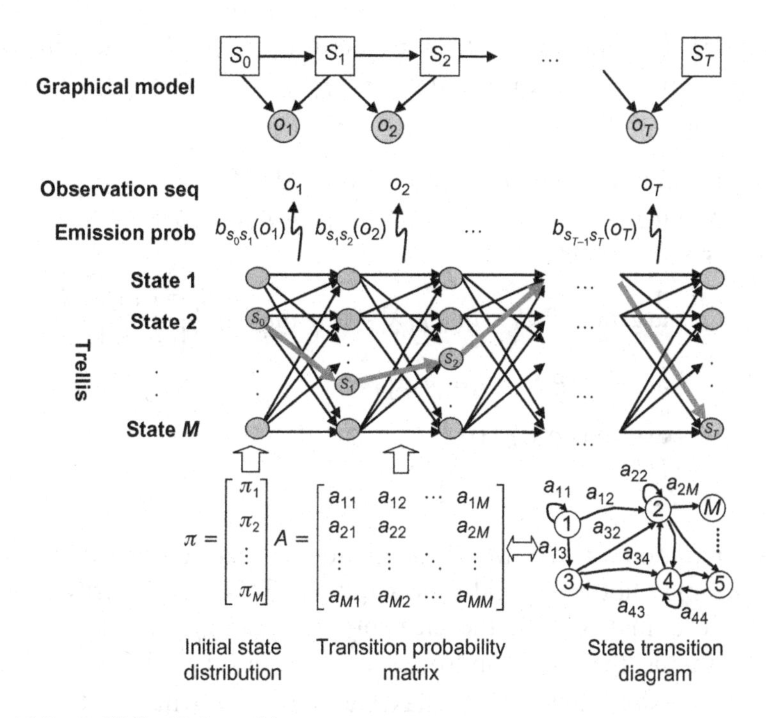

Figure 1.3 Standard hidden Markov model.
In the graphical model, *an HMM has one discrete hidden node and one discrete or continuous observed node per slice. Circles denote continuous nodes, squares denote discrete nodes, clear means hidden, shaded means observed. The arc from node U to node V indicates that U "causes" V. An instance of the hidden Markov process is shown in the* trellis, *where the thick line represents a state path, and the thin lines represent the available transitions of states. The state transition probabilities are specified in the* transition probability matrix A *with the initial state* s_0 *selected according to the* initial state distribution π. *Equivalent to the state transition probability matrix* A, *the underlying Markov chain of the HMM can be expressed by the* state transition diagram. *The process produces observation* o_1 *with emission probability* $b_{s_0,s_1}(o_1)$ *while transiting from state* s_0 *to* s_1, o_2 *with* $b_{s_1,s_2}(o_2)$ *while from* s_1 *to* $s_2, \ldots,$ *until the final observation* o_T.

on the current state and independent of the partial observed sequence $o_{1:t}$ given the current state is applied, that is,

$$P[o_{t+1:T}|S_t = j, o_{1:t}, \lambda] = P[o_{t+1:T}|S_t = j, \lambda].$$

Then the *forward–backward algorithm* (also called *Baum–Welch algorithm*) for HMM is

$$\begin{aligned} \alpha_t(j) &= \sum_{i \in \mathbf{S}} P[S_{t-1} = i, S_t = j, o_{1:t}|\lambda] \\ &= \sum_{i \in \mathbf{S}} P[S_{t-1} = i, o_{1:t-1}|\lambda] P[S_t = j, o_t|S_{t-1} = i, \lambda] \\ &= \sum_{i \in \mathbf{S}} \alpha_{t-1}(i) a_{ij} b_{ij}(o_t), \quad 1 \leq t \leq T, j \in \mathbf{S}, \end{aligned} \tag{1.1}$$

$$\beta_t(j) = \sum_{i \in \mathbf{S}} a_{ji} b_{ji}(o_{t+1}) \beta_{t+1}(i), \quad 0 \leq t \leq T-1, j \in \mathbf{S}, \tag{1.2}$$

with the initial conditions $\alpha_0(j) = \pi_j$ and $\beta_T(j) = 1$, $j \in \mathbf{S}$, where in order to make Eqn (1.2) true for $\beta_{T-1}(j) = P[o_T | S_{T-1} = j, \lambda] = \sum_{i \in \mathbf{S}} a_{ji} b_{ji}(o_T)$, it is assumed $\beta_T(j) = 1$.

The maximum *a posteriori* probability (MAP) estimation $\hat{s}_t$ of state S_t at a given time t after the entire sequence $o_{1:T}$ has been observed can be determined by

$$\begin{aligned}
\hat{s}_t &= \arg\max_{i \in \mathbf{S}} P[S_t = i | o_{1:T}, \lambda] = \arg\max_{i \in \mathbf{S}} \frac{P[S_t = i, o_{1:T} | \lambda]}{P[o_{1:T} | \lambda]} \\
&= \arg\max_{i \in \mathbf{S}} P[S_t = i, o_{1:T} | \lambda] \\
&= \arg\max_{i \in \mathbf{S}} \alpha_t(i) \beta_t(i),
\end{aligned}$$

for $0 \leq t \leq T$.

A limitation of the MAP estimation is that the resulting state sequence may not be a valid state sequence. This can be easily proved. Let $\mathbf{C}$ be a matrix with the elements $c_{ij} = \alpha_t(i) a_{ij} b_{ij}(o_{t+1}) \beta_{t+1}(j)$ and $c_{i'j'} = 0$, that is, the transition from state i' to j' at time t is not valid. But it is still possible that $i' = \arg\max_i \mathbf{Ce}$ and $j' = \arg\max_j \mathbf{e}^T \mathbf{C}$, that is, those two states could be the best choice based on the MAP estimation, where $\mathbf{e}$ is an all-ones vector and $\mathbf{e}^T$ its transpose. However, this limitation does not affect its successful application in vast areas including the area of digital communications. A famous algorithm in the digital communication area is the *BCJR algorithm* (Bahl et al., 1974), which replaces $b_{ij}(o_t)$ with $\sum_{x_t} e_{ij}(x_t) R(o_t, x_t)$ in the forward–backward formulas (1.1) and (1.2) for decoding, where x_t is the output associated with a state transition at time t, $e_{ij}(x_t) = P[x_t | S_{t-1} = i, S_t = j, \lambda]$ is the output probability, o_t is the observation of x_t and $R(o_t, x_t) = P[o_t | x_t]$ is the channel transition probability. The difference between the Baum–Welch algorithm and the BCJR algorithm is shown in Figure 1.4.

Similar to the derivation of the forward formula for the HMM, the Viterbi algorithm can be readily derived. Define

$$\delta_t(j) \equiv \max_{S_{1:t-1}} P[S_{1:t-1}, S_t = j, o_{1:t} | \lambda].$$

Then from the similarity we can see that by replacing the sum–product of Eqn (1.1) with the max-product the *Viterbi algorithm* of HMM is yielded by:

$$\delta_t(j) = \max_{i \in \mathbf{S}} \{\delta_{t-1}(i) a_{ij} b_{ij}(o_t)\}, \quad 1 \leq t \leq T, \ j \in \mathbf{S},$$

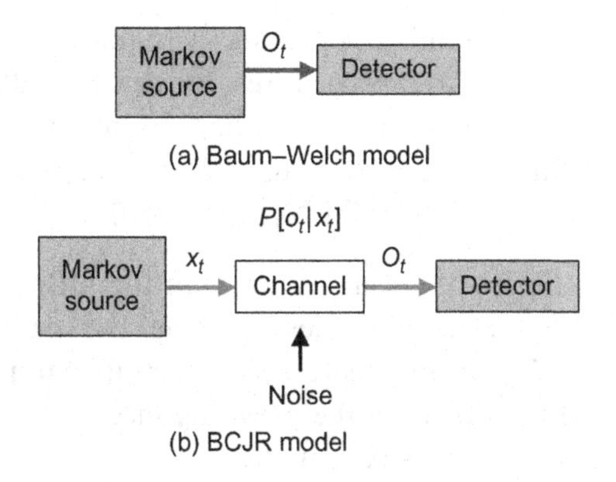

Figure 1.4 Baum–Welch and BCJR models.

(a) The hidden Markov source produces signal o_t at time t, for $t = 1, \ldots, T$, according to the emission probabilities $b_{s_{t-1},s_t}(o_t)$ while the source transits from state s_{t-1} to s_t. The sequence of signals is detected by the detector, which is used to estimate the states of the source.

(b) The hidden Markov source produces signal x_t at time t, for $t = 1, \ldots, T$, according to the emission probabilities $e_{s_{t-1},s_t}(x_t)$ while the source transits from state s_{t-1} to s_t. Due to noise of the channel, the signal x_t is transformed to o_t according to the transition probability $P[o_t|x_t]$ while transferring over the channel. The sequence $o_{1:T}$ observed by the detector is used to estimate the states of the source.

where the initial value $\delta_0(j) = \pi_j$, and $S_{1:t-1} = (S_1, \ldots, S_{t-1})$ is the partial state sequence. The variable $\delta_t(j)$ represents the score of the *surviving sequence* among all possible sequences that enter the state j at time t. Therefore, by tracing back from $\max_{j \in S}\delta_T(j)$ one can find the *optimal state sequence* $(S_1^*, \ldots, S_T^*)$ that maximizes the *likelihood function* $P[o_{1:T}|S_{1:T}, \lambda]$, when the prior state sequence $S_{1:T}$ is uniform.

1.3 DYNAMIC BAYESIAN NETWORKS

A BN is defined by a directed acyclic graph (DAG) in which nodes correspond to random variables having conditional dependences on the parent nodes, and arcs represent the dependencies between variables. A dynamic Bayesian networks (DBN) was first proposed by Dean and Kanazawa (1989), which extends the BN by providing an explicit discrete temporal dimension that uses arcs to establish dependencies between variables in different time slices. DBNs can be generally used to represent very complex temporal processes. Various HMMs and hidden semi-Markov models (HSMMs) as well as state space models (SSMs) can be expressed using the DBN models (Murphy, 2002b). For example, an HMM can be represented as a DBN with a single state variable and a single observation variable in each slice, as shown in Figure 1.3. In contrast, each slice of a general

DBN can have any number of state variables and observation variables. Because there are procedures for learning the parameters and structures, DBNs have been applied in many areas, such as motion processes in computer vision, speech recognition, genomics, and robot localization (Russell and Norvig, 2010).

Let $\mathbf{U}_t$ denote the set of variables/nodes in the DBN at time slice t, and $X_t \in \mathbf{U}_t$ be one of its variables. Denote the parents of X_t by $Pa(X_t) = Pa_t(X_t) \cup Pa_{t-1}(X_t)$, where $Pa_t(X_t)$ are its parents in the same slice and $Pa_{t-1}(X_t)$ the others in the previous slice. Let $\mathbf{U}_t^{(k)}$ be a subset of $\mathbf{U}_t$, and define the set of its parents by

$$Pa(\mathbf{U}_t^{(k)}) = \{Pa(X_t) : X_t \in \mathbf{U}_t^{(k)}\}.$$

Similarly define $Pa_t(\mathbf{U}_t^{(k)})$ and $Pa_{t-1}(\mathbf{U}_t^{(k)})$ such that $Pa(\mathbf{U}_t^{(k)}) = Pa_t(\mathbf{U}_t^{(k)}) \cup Pa_{t-1}(\mathbf{U}_t^{(k)})$.

Because the DBN is a DAG, $\mathbf{U}_t$ can be divided into non-overlapped subsets $\mathbf{U}_t^{(1)}, \ldots, \mathbf{U}_t^{(K)}$ such that for each $X_t^{(k+1)} \in \mathbf{U}_t^{(k+1)}$, there exist $Pa_t(X_t^{(k+1)}) \subseteq \bigcup_{l=1}^{k} \mathbf{U}_t^{(l)}$ and $Pa_t(X_t^{(k+1)}) \cap \mathbf{U}_t^{(k)} \neq null$. That is, the intra-slice parents of each $X_t^{(k+1)} \in \mathbf{U}_t^{(k+1)}$ belong to the previous subsets $\mathbf{U}_t^{(1)}, \ldots, \mathbf{U}_t^{(k)}$, and it has at least one parent belonging to $\mathbf{U}_t^{(k)}$.

Based on this definition, the start subset $\mathbf{U}_t^{(1)}$ has no intra-slice parents, that is, $Pa_t(\mathbf{U}_t^{(1)}) = null$. Each variable $X_t^{(2)} \in \mathbf{U}_t^{(2)}$ has at least one parent belonging to $\mathbf{U}_t^{(1)}$. Therefore, there exists at least one directed path of length 1 from $\mathbf{U}_t^{(1)}$ to $X_t^{(2)}$. Hence, we can conclude that for each $X_t^{(k+1)} \in \mathbf{U}_t^{(k+1)}$, there exists at least one directed path of length k that across $\mathbf{U}_t^{(1)}, \mathbf{U}_t^{(2)}, \ldots, \mathbf{U}_t^{(k)}$, until $X_t^{(k+1)}$. But this does not prevent $X_t^{(k+1)}$ from having a shorter path starting from any of the previous subsets. In other words, the maximum length of directed path to $X_t^{(k+1)} \in \mathbf{U}_t^{(k+1)}$ must be k. The last subset $\mathbf{U}_t^{(K)}$ contains the variables that have the longest paths of length K, and has no intra-slice children. Any subset $\mathbf{U}_t^{(k)}$, $k = 1, \ldots, K$, can have inter-slice parents, that is, $Pa_{t-1}(\mathbf{U}_t^{(k)}) \neq null$.

As shown in Figure 1.5, a general DBN can be expressed by a left-to-right network, in which each "node" represents a subset of nodes of

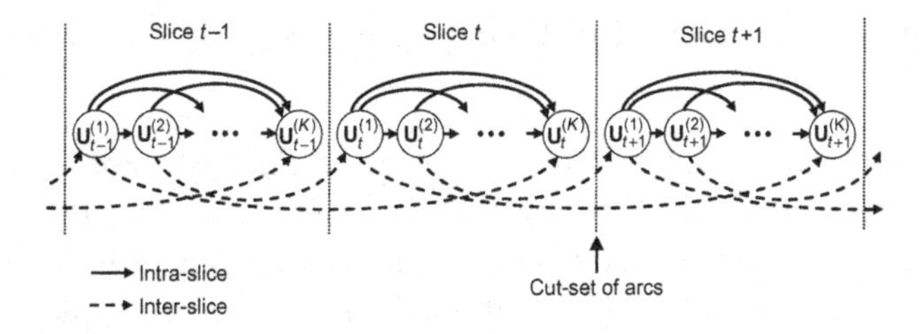

Figure 1.5 Left-right expression of a general DBN.
Each of the circles represents a subset of nodes of the DBN, and the arcs represent the dependency that a node in the child subset is dependent on its parents in the parent subset. All the slices, except the first and the last slices (not shown in the figure), have the same structure of DBN. In slice t*, the first subset of nodes,* $\mathbf{U}_t^{(1)}$*, has inter-slice parents in, for instance,* $\mathbf{U}_{t-1}^{(2)}$ *of the previous slice* t − 1 *without intra-slice parents. The other subsets of nodes in slice* t *may have both inter-slice and intra-slice parents. For example, the subset of nodes,* $\mathbf{U}_t^{(K)}$*, has inter-slice parents in* $\mathbf{U}_{t-1}^{(1)}$ *and intra-slice parents in* $\mathbf{U}_t^{(1)}$*,* $\mathbf{U}_t^{(2)}$*, and* $\mathbf{U}_t^{(K-1)}$*. In contrast, the subset of nodes,* $\mathbf{U}_t^{(2)}$*, has no inter-slice parents in the previous slice* t − 1*. Besides, each subset of nodes,* $\mathbf{U}_t^{(k)}$*, must have at least one parent in* $\mathbf{U}_t^{(k-1)}$*, for* k = 2, …, K.

the original DBN and each "arc" denotes that the child subset has at least one parent in the parent subset. The solid arrows represent the intra-slice arcs and the dotted arrows the inter-slice arcs.

Based on the partition of DAG, the transition probabilities from $\mathbf{U}_{t-1}$ to $\mathbf{U}_t$ are:

$$\begin{aligned} P[\mathbf{U}_t|\mathbf{U}_{t-1}] &= P[\mathbf{U}_t^{(1)}|\mathbf{U}_{t-1}]\cdots P\left[\mathbf{U}_t^{(K)}|\mathbf{U}_{t-1}, \bigcup_{k=1}^{K-1}\mathbf{U}_t^{(k)}\right] \\ &= \prod_{X_t^{(1)}\in\mathbf{U}_t^{(1)}} P[X_t^{(1)}|Pa_{t-1}(X_t^{(1)})]\cdots \\ &\quad \prod_{X_t^{(K)}\in\mathbf{U}_t^{(K)}} P\left[X_t^{(K)}|Pa_{t-1}(X_t^{(K)}), Pa_t(X_t^{(K)})\right] \\ &= \prod_{X_t\in\mathbf{U}_t} P[X_t|Pa(X_t)]. \end{aligned} \tag{1.3}$$

By unrolling the network until T slices, the joint distribution for a sequence of length T can be obtained:

$$P[\mathbf{U}_{1:T}] = P[\mathbf{U}_0]\cdot\prod_{t=1}^{T} P[\mathbf{U}_t|\mathbf{U}_{t-1}],$$

where $P[\mathbf{U}_0]$ is the initial distribution of the random variables. Let $\mathbf{u}_t = \{X_t = x_t : X_t\in\mathbf{U}_t\}$ be a realization of $\mathbf{U}_t$. Then $P[\mathbf{u}_t]$ can be calculated recursively by

$$P[\mathbf{u}_t] = \sum_{Pa_{t-1}(\mathbf{u}_t)} P[Pa_{t-1}(\mathbf{u}_t)]P[\mathbf{u}_t|Pa_{t-1}(\mathbf{u}_t)], \tag{1.4}$$

where $Pa_{t-1}(\mathbf{u}_t)$ are the set of boundary nodes going out of slice $t-1$ to t.

For each realization x_t and an instance $Pa(x_t)$, there is a model parameter $P[x_t|Pa(x_t)]$. Then by Eqn (1.3) the transition probabilities $P[\mathbf{u}_t|\mathbf{u}_{t-1}]$ can be determined with N multiplications, where N is the number of random variables contained in $\mathbf{U}_t$. For example, if every random variable has D parents and can take M discrete values, then for a variable X_t the DBN requires M^{D+1} parameters to represent the transition probabilities $P[X_t|Pa(X_t)]$. Hence, the DBN requires NM^{D+1} parameters to represent the transition probabilities $P[\mathbf{U}_t|\mathbf{U}_{t-1}]$.

Denote $\mathbf{U}_t^{(E)}$ as the set of entry nodes from slice $t-1$ into slice t by letting

$$\mathbf{U}_t^{(E)} \equiv \{X_t : Pa_{t-1}(X_t) \neq null\}.$$

Let $\mathbf{U}_t^{(I)} = \mathbf{U}_t \backslash \mathbf{U}_t^{(E)}$ be the set of inner nodes of slice t. Then for an inner node X_t of $\mathbf{U}_t^{(k+1)}$, that is, $X_t \in \mathbf{U}_t^{(I)} \cap \mathbf{U}_t^{(k+1)}$, we have $Pa_{t-1}(X_t) = null$ and

$$Pa(X_t) = Pa_t(X_t) \subseteq \bigcup_{l=1}^{k} \mathbf{U}_t^{(l)} \subseteq \mathbf{U}_t^{(E)} \cup \bigcup_{l=1}^{k} \left(\mathbf{U}_t^{(I)} \cap \mathbf{U}_t^{(l)}\right).$$

Therefore,

$$\begin{aligned}
P[\mathbf{U}_t^{(I)}|\mathbf{U}_t^{(E)}] &= P\left[\bigcup_{k=1}^{K}\left(\mathbf{U}_t^{(I)} \cap \mathbf{U}_t^{(k)}\right)|\mathbf{U}_t^{(E)}\right] \\
&= P\left[\left(\mathbf{U}_t^{(I)} \cap \mathbf{U}_t^{(1)}\right)|\mathbf{U}_t^{(E)}\right] \cdot \\
&\quad P\left[\bigcup_{k=2}^{K}\left(\mathbf{U}_t^{(I)} \cap \mathbf{U}_t^{(k)}\right)|\mathbf{U}_t^{(E)} \cup \left(\mathbf{U}_t^{(I)} \cap \mathbf{U}_t^{(1)}\right)\right] \\
&\ldots \\
&= \prod_{k=1}^{K} P\left[\mathbf{U}_t^{(I)} \cap \mathbf{U}_t^{(k)}|\mathbf{U}_t^{(E)} \cup \sum_{l=1}^{k-1} \mathbf{U}_t^{(I)} \cap \mathbf{U}_t^{(l)}\right] \\
&= \prod_{k=1}^{K} \prod_{X_t \in \mathbf{U}_t^{(I)} \cap \mathbf{U}_t^{(k)}} P[X_t|Pa(X_t)] \\
&= \prod_{X_t \in \mathbf{U}_t^{(I)}} P[X_t|Pa(X_t)].
\end{aligned}$$

That is, $P[\mathbf{u}_t^{(I)}|\mathbf{u}_t^{(E)}]$ can be determined by $|\mathbf{U}_t^{(I)}|$ multiplications of the corresponding model parameters. This implies that the transition from slice $t-1$ to slice t has two stages: inter-slice transition from slice $t-1$ to slice t with probability $P[\mathbf{u}_t^{(E)}|\mathbf{u}_{t-1}]$, and intra-slice transition within slice t with probability $P[\mathbf{u}_t^{(I)}|\mathbf{u}_t^{(E)}]$.

Because $Pa_t(\mathbf{U}_t^{(k+1)}) \subseteq \bigcup_{l=1}^{k} \mathbf{U}_t^{(l)}$ according to the definition of $\mathbf{U}_t^{(k+1)}$, we have $Pa_t(\mathbf{U}_t^{(k+1)}) \cap \mathbf{U}_t^{(k+1)} = null$ and $Pa_t\left(\bigcup_{l=1}^{k} \mathbf{U}_t^{(l)}\right) \cap \mathbf{U}_t^{(k+1)} = null$. In other words, any variable $X_t \in \mathbf{U}_t^{(k+1)}$ cannot be a parent of $\mathbf{U}_t^{(1)}, \ldots, \mathbf{U}_t^{(k+1)}$. Therefore, the transition probabilities $P[\mathbf{u}_t^{(I)}|\mathbf{u}_t^{(E)}]$ for given $\mathbf{u}_t^{(E)}$ satisfy:

$$
\begin{aligned}
\sum_{\mathbf{u}_t^{(I)}} P[\mathbf{u}_t^{(I)}|\mathbf{u}_t^{(E)}] &= \sum_{\mathbf{u}_t^{(I)}} \prod_{k=1}^{K} \prod_{x_t \in \mathbf{u}_t^{(I)} \cap \mathbf{u}_t^{(k)}} P[x_t|Pa(x_t)] \\
&= \sum_{\mathbf{u}_t^{(I)} \backslash \mathbf{u}_t^{(K)}} \prod_{k=1}^{K-1} \prod_{x_t \in \mathbf{u}_t^{(I)} \cap \mathbf{u}_t^{(k)}} P[x_t|Pa(x_t)] \sum_{\mathbf{u}_t^{(K)}} \prod_{x_t' \in \mathbf{u}_t^{(I)} \cap \mathbf{u}_t^{(K)}} P[x_t'|Pa(x_t')] \\
&= \sum_{\mathbf{u}_t^{(I)} \backslash \mathbf{u}_t^{(K)}} \prod_{k=1}^{K-1} \prod_{x_t \in \mathbf{u}_t^{(I)} \cap \mathbf{u}_t^{(k)}} P[x_t|Pa(x_t)] \\
&\cdots \\
&= 1.
\end{aligned}
$$

Since $P[\mathbf{U}_t^{(I)}|\mathbf{U}_t^{(E)}, \mathbf{U}_{t-1}] = P[\mathbf{U}_t^{(I)}|\mathbf{U}_t^{(E)}]$, we have a forward formula

$$
\begin{aligned}
P[\mathbf{u}_t^{(E)}] &= \sum_{\mathbf{u}_{t-1}^{(E)}, \mathbf{u}_{t-1}^{(I)}, \mathbf{u}_t^{(I)}} P[\mathbf{u}_{t-1}^{(E)}, \mathbf{u}_{t-1}^{(I)}, \mathbf{u}_t^{(E)}, \mathbf{u}_t^{(I)}] \\
&= \sum_{\mathbf{u}_{t-1}^{(E)}, \mathbf{u}_{t-1}^{(I)}, \mathbf{u}_t^{(I)}} P[\mathbf{u}_{t-1}^{(E)}] P[\mathbf{u}_{t-1}^{(I)}|\mathbf{u}_{t-1}^{(E)}] P[\mathbf{u}_t^{(E)}|\mathbf{u}_{t-1}] P[\mathbf{u}_t^{(I)}|\mathbf{u}_t^{(E)}], \\
&= \sum_{\mathbf{u}_{t-1}^{(E)}, \mathbf{u}_{t-1}^{(I)}} P[\mathbf{u}_{t-1}^{(E)}] P[\mathbf{u}_{t-1}^{(I)}|\mathbf{u}_{t-1}^{(E)}] P[\mathbf{u}_t^{(E)}|\mathbf{u}_{t-1}^{(D)}]
\end{aligned}
\tag{1.5}
$$

where $\mathbf{u}_{t-1}^{(D)} \equiv Pa_{t-1}(\mathbf{u}_t^{(E)})$ are the set of boundary nodes going out of slice $t-1$ to t. Then $P[\mathbf{u}_t] = P[\mathbf{u}_t^{(E)}]P[\mathbf{u}_t^{(I)}|\mathbf{u}_t^{(E)}]$. It can be seen that when $|\mathbf{U}_t^{(E)}| < N$ and $P[\mathbf{u}_t^{(E)}|\mathbf{u}_{t-1}^{(D)}]$ can be easily obtained, Eqn (1.5) has fewer dimensions than Eqn (1.4).

As shown in Figure 1.5, any cut-set of arcs divides the DBN into a left part and a right part. If the starting/ending nodes of the arcs in the cut-set are given, such as given $\mathbf{u}_{t-1}^{(D)}$ or $\mathbf{u}_t^{(E)}$, the transition probabilities to the nodes in the right part can be completely determined. For instance, for given $\mathbf{u}_t^{(E)}$, the transition probabilities $P[\mathbf{u}_t^{(I)}|\mathbf{u}_t^{(E)}]$ can be determined and $\sum_{\mathbf{u}_t^{(I)}} P[\mathbf{u}_t^{(I)}|\mathbf{u}_t^{(E)}] = 1$. Based on this observation, we can try to find a cut-set of arcs, which can be inter-slice, intra-slice, or mixed ones, so that they have the minimum number of starting/ending nodes. Let $\mathbf{U}_t^{(Min)}$ denote the set of nodes in one ends of the cut-set of arcs. Then $P[\mathbf{U}_t^{(Min)}]$ may have the minimum number of dimensions.

Suppose one random variable in $\mathbf{U}_t$ is the observation o_t and the others are states. Denote the set of states by $\mathbf{S}_t$. The observation o_t is given and cannot be a parent of a state. The observation probabilities are $P[o_t|Pa(o_t)]$. Therefore, if we define the forward variables by

$$\begin{aligned}\alpha_t(\mathbf{s}_t) &\equiv P[o_{1:t}, \mathbf{s}_t] = \sum_{\mathbf{s}_{t-1}} P[o_{1:t}, \mathbf{s}_{t-1}, \mathbf{s}_t] \\ &= \sum_{\mathbf{s}_{t-1}} \alpha_{t-1}(\mathbf{s}_{t-1}) P[\mathbf{s}_t|\mathbf{s}_{t-1}] P[o_t|o_{1:t-1}, \mathbf{s}_{t-1}, \mathbf{s}_t] \\ &= \sum_{\mathbf{s}_{t-1}} \alpha_{t-1}(\mathbf{s}_{t-1}) P[\mathbf{s}_t|\mathbf{s}_{t-1}^{(D)}] P[o_t|Pa(o_t)]\end{aligned} \tag{1.6}$$

then the DBN becomes a HMM. Similarly, the forward variables can be defined with reduced-dimensions by

$$\begin{aligned}\alpha_t(\mathbf{s}_t^{(E)}) &\equiv P[o_{1:t}, \mathbf{s}_t^{(E)}] \\ &= \sum_{\mathbf{s}_{t-1}^{(E)}, \mathbf{s}_{t-1}^{(I)}, \mathbf{s}_t^{(I)}} \alpha_{t-1}(\mathbf{s}_{t-1}^{(E)}) P[\mathbf{s}_{t-1}^{(I)}|\mathbf{s}_{t-1}^{(E)}] P[\mathbf{s}_t^{(E)}, \mathbf{s}_t^{(I)}|\mathbf{s}_{t-1}^{(D)}] P[o_t|Pa(o_t)] \\ &= \sum_{\mathbf{s}_{t-1}^{(E)}, \mathbf{s}_{t-1}^{(I)}, \mathbf{s}_t^{(I)}} \alpha_{t-1}(\mathbf{s}_{t-1}^{(E)}) P[\mathbf{s}_{t-1}^{(I)}|\mathbf{s}_{t-1}^{(E)}] P[\mathbf{s}_t^{(E)}|\mathbf{s}_{t-1}^{(D)}] P[\mathbf{s}_t^{(I)}|\mathbf{s}_t^{(E)}] P[o_t|Pa(o_t)].\end{aligned} \tag{1.7}$$

The backward variables for HMM can be defined by $\beta_t(\mathbf{s}_t) \equiv P[o_{t+1:T}|\mathbf{s}_t]$ or $\beta_t(\mathbf{s}_t^{(E)}) \equiv P[o_{t+1:T}|\mathbf{s}_t^{(E)}]$, and so the backward formula can be readily derived.

Compared with Eqns (1.4) and (1.5), we can see that in the discrete cases the computational complexity of Eqns (1.6) and (1.7) and the number of model parameters for an HMM are consistent to the corresponding DBN. In the literature, one generally argues that the HMM would require M^{2N} model parameters to represent the transition probabilities $P[\mathbf{U}_t|\mathbf{U}_{t-1}]$ and thus would be much complex than the corresponding DBN model while every random variable has few parents.

Using the forward–backward algorithms of HMMs, the model parameters of a DBN can be estimated. Based on the given set of model parameters λ, a variety of inference problems can be solved including: filtering $P[\mathbf{S}_t|o_{1:t}, \lambda]$, predicting $P[\mathbf{S}_{t+\tau}|o_{1:t}, \lambda]$ for $\tau > 0$, fixed-lag smoothing $P[\mathbf{S}_{t-\tau}|o_{1:t}, \lambda]$, fixed-interval smoothing $P[\mathbf{S}_t|o_{1:T}, \lambda]$ for given observation sequence, most likely path finding $\arg\max_{\mathbf{S}_{1:t}} P[\mathbf{S}_{1:t}|o_{1:t}, \lambda]$, and likelihood computing $P[o_{1:t}|\lambda] = \sum_{\mathbf{S}_{1:t}} P[\mathbf{S}_{1:t}, o_{1:t}|\lambda]$. Therefore, for a discrete-state DBN the simplest inference method is to apply the forward–backward algorithms of HMMs. Though it turns out that the "constant" for the per-update time and space, the complexity is almost always exponential in the number of state variables.

For exact inference in DBNs, another simple method is unrolling the DBNs until it accommodates the whole sequence of observations and then using any algorithm for inference in BNs such as variable elimination, clustering methods, etc. However, a naive application of unrolling would not be particularly efficient for a long sequence of observations.

In general, one can use DBNs to represent very complex temporal processes. However, even in the cases that have sparsely connected variables, one cannot reason efficiently and exactly about those processes. Applying approximate methods seems to be the way. Among a variety of approximation methods, the particle filtering (Gordon et al., 1993) is a sequential importance re-sampling method. It is very commonly used to estimate the posterior density of the state variables given the observation variables. Particle filtering is consistent and efficient, and can be used for discrete, continuous, and hybrid DBNs.

1.4 CONDITIONAL RANDOM FIELDS

Different from the directed graphical model of DBNs, conditional random fields (CRFs) are a type of undirected probabilistic graphical model whose nodes can be divided into exactly two disjoint sets, that is, the observations **O** and states **S**. Therefore, a DBN is a model of the joint distribution $P[\mathbf{O}, \mathbf{S}]$, while a CRF is a model of the conditional distribution $P[\mathbf{S}|\mathbf{O}]$ because **O** is not necessary to be included into the same graphical structure as **S**. Hence, in a CRF, only **S** is assumed being indexed by the vertices of an undirected graph $G = (V,E)$, and is globally conditioned on the observations **O**. The rich and global features of the observations can then be used, while dependencies among the observations do not need to be explicitly represented. In the literature, the directed graphical model of DBNs is referred as a generative model in which the observations **O** usually cannot be a parent of a state and be considered being probabilistically generated by the states **S**. In contrast, the undirected graphical model is termed as a discriminative model in which a model of $P[\mathbf{O}]$ is not required. CRFs have been applied in shallow parsing, named entity recognition, gene finding, object recognition, and image segmentation (Sutton and McCallum, 2006).

A CRF is generally defined by $P[\mathbf{s}|\mathbf{o}]$ if for any fixed **o**, the distribution $P[\mathbf{s}|\mathbf{o}]$ factorizes according to a factor graph G of **S** (Sutton and McCallum, 2006). If the factor graph $G = (V, E)$ of **S** is a chain or a tree, the conditional distribution $P[\mathbf{s}|\mathbf{o}]$ has the form

$$P[\mathbf{s}|\mathbf{o}] = \frac{1}{Z(\mathbf{o})}\exp\left(\sum_{e\in E,k}\lambda_k f_k(e, \mathbf{s}_e, \mathbf{o}) + \sum_{v\in V,k'}\mu_{k'} g_{k'}(v, \mathbf{s}_v, \mathbf{o})\right), \quad (1.8)$$

where $\mathbf{s}_e$ and $\mathbf{s}_v$ are the component sets of **s** associated with the vertices of edge e and vertice v, respectively, the feature functions f_k and $g_{k'}$ are given and fixed, and $Z(\mathbf{o})$ is an instance-specific normalization function defined by

$$Z(\mathbf{o}) = \sum_{\mathbf{s}}\exp\left(\sum_{e\in E,k}\lambda_k f_k(e, \mathbf{s}_e, \mathbf{o}) + \sum_{v\in V,k'}\mu_{k'} g_{k'}(v, \mathbf{s}_v, \mathbf{o})\right). \quad (1.9)$$

Example 1.3 From an HMM to a CRF

For an instance of observation sequence $\mathbf{o} = (o_1, \ldots, o_T)$ and state sequence $\mathbf{s} = (s_1, \ldots, s_T)$, the set of vertices of the factor graph G is $V = \{s_1, \ldots, s_T\}$, and the set of edges is $E = \{s_{t-1}, s_t : t = 2, \ldots, T\}$, where s_1 is assumed being the initial state. Then the conditional probability factorizes as

$$
\begin{aligned}
P[\mathbf{s}|\mathbf{o}] &= \frac{1}{P[\mathbf{o}]}P[\mathbf{s},\mathbf{o}] = \frac{1}{P[\mathbf{o}]}\prod_{t=1}^{T}P[s_t|s_{t-1}]P[o_t|s_t] \\
&= \frac{1}{P[\mathbf{o}]}\exp\left(\sum_{t=1}^{T}\log P[s_t|s_{t-1}] + \sum_{t=1}^{T}\log P[o_t|s_t]\right) \\
&= \frac{1}{P[\mathbf{o}]}\exp\left(\sum_{t=1}^{T}\sum_{i,j}\log a_{ij}\cdot I((s_{t-1},s_t)=(i,j))\right. \\
&\quad \left. + \sum_{t=1}^{T}\sum_{j,x}\log b_j(x)\cdot I(s_t=j)\cdot I(o_t=x)\right),
\end{aligned}
$$

where $P[s_1]$ is represented by $P[s_1|s_0]$ for simplicity, and $I(x=x')$ denotes an indicator function of x which takes the value 1 when $x=x'$ and 0 otherwise. Let $Z(\mathbf{o})=P[\mathbf{o}]$, $\lambda_k=\log a_{ij}$, $k\leftrightarrow(i,j)$, $f_k(e=(s_{t-1},s_t),\mathbf{s}_e,\mathbf{o})=I((s_{t-1},s_t)=(i,j))$, $\mu_{k'}=\log b_j(x)$, $k'\leftrightarrow(j,x)$, and $g_{k'}(v=s_t,\mathbf{s}_v,\mathbf{o})=I(s_t=j)\cdot I(o_t=x)$. Then the HMM becomes a linear-chain CRF, as given by Eqn (1.8). $Z(\mathbf{o})$ is determined by Eqn (1.9).

From this example, we can see that an HMM can be expressed by a CRF with specific feature functions. CRFs extend the HMM to containing any number of feature functions that can inspect the entire observation sequence. The feature functions f_k and $g_{k'}$ need not have a probabilistic interpretation.

When the graph G is a chain or a tree, the algorithms analogous to the forward–backward algorithms, and Viterbi algorithm for HMMs can be used to yield exact inference in a CRF. However, for a general graph, the problem of exact inference in a CRF is intractable, and algorithms, such as loopy belief propagation, can be used to obtain approximate solutions.

To estimate the parameters $\theta=\{\lambda_k\}\cup\{\mu_{k'}\}$ using training data $\mathbf{o}^{(1)},\ldots,\mathbf{o}^{(N)}$ and $\mathbf{s}^{(1)},\ldots,\mathbf{s}^{(N)}$, maximum likelihood learning is used, where $\mathbf{o}^{(i)}=(o_1^{(i)},\ldots,o_T^{(i)})$ is the ith observation sequence and $\mathbf{s}^{(i)}=(s_1^{(i)},\ldots,s_T^{(i)})$ is the corresponding ith state sequence. For the CRFs expressed by Eqn (1.8) where all nodes have exponential family distributions and $\mathbf{s}^{(1)},\ldots,\mathbf{s}^{(N)}$ observe all nodes of the graph, this optimization is convex. Gradient descent algorithms, Quasi-Newton methods, such as the L-BFGS algorithm, can be used to solve the problem. In the case that some states are unobserved, those states have to be inferred.

1.5 HIDDEN SEMI-MARKOV MODELS

Due to the non-zero probability of self-transition of a nonabsorbing state, the state duration of an HMM is implicitly a geometric distribution. This makes the HMM has limitations in some applications. A HSMM is traditionally defined by allowing the underlying process to be a semi-Markov chain. Each state has a variable duration, which is associated with the number of observations produced while in the state. The HSMM is also called explicit duration HMM (Ferguson, 1980; Rabiner, 1989), variable-duration HMM (Levinson, 1986a; Russell and Moore, 1985; Rabiner, 1989), HMM with explicit duration (Mitchell et al., 1995), HSMM (Murphy, 2002a), generalized HMM (Kulp et al., 1996), segmental HMM (Russell, 1993), and segment model (Ostendorf and Roukos, 1989; Ostendorf et al., 1996) in the literature, depending on their assumptions and their application areas.

In the simplest case when the observations are conditionally independent of each other for given states, an HSMM can be expressed using an HMM, DBN, or CRF. However, a general HSMM cannot be expressed by a fixedly structured graphical model because the number of nodes in the graphical model is itself random which in turn emits random-length segments of observations. For example, for a given length T of observation sequence $(o_1, \ldots, o_T)$, the length $N \le T$ of the state sequence $(S_1, \ldots, S_N)$ is a random variable due to the variable state durations. We do not know in advance how to divide the observation sequence into N segments corresponding to the state sequence, and let o_t be conditioned on which state variable.

A general HSMM is shown in Figure 1.6. In the figure, the actual sequence of events is taken to be:

1. The first state i_1 and its duration d_1 are selected according to the state transition probability $a_{(i_0,d_0)(i_1,d_1)}$, where (i_0, d_0) is the initial state and duration. State i_1 lasts for $d_1 = 2$ time units in this instance.
2. It produces two observations (o_1, o_2) according to the emission probability $b_{i_1,d_1}(o_1, o_2)$.

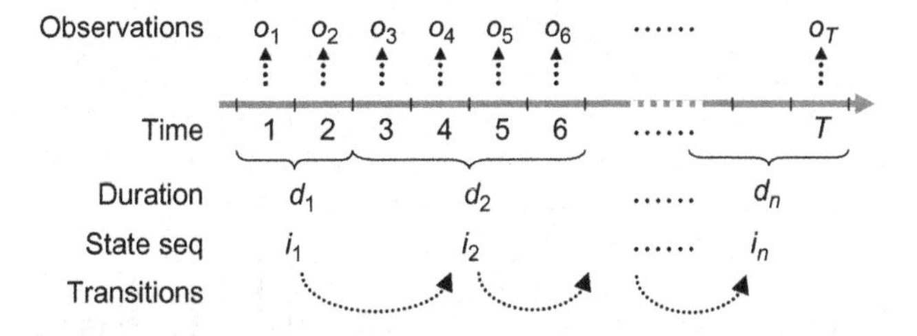

Figure 1.6 A general HSMM.
The time is discrete. At each time interval, there is an observation, which is produced by the hidden/unobservable state. Each state lasts for a number of time intervals. For example, state i_2 lasts for 4 time intervals, producing 4 observations: o_3, o_4, o_5, and o_6. When a state ends at time t, it transits to the next state at time t + 1. For example, state i_1 ends at time 2 and transits to i_2 at time 3. The transition is instantaneously finished at the jump time from time 2 to 3. The state sequence (i_1, d_1), ..., (i_n, d_n) is corresponding to the observation sequence (o_1, ..., o_T) with $\sum_{m=1}^{n} d_m = T$, where T is the length of the observation sequence, and n is the number of state jumps/transitions (including from initial state to the first state).

3. It transits, according to the state transition probability $a_{(i_1,d_1)(i_2,d_2)}$, to state i_2 with duration d_2.
4. State i_2 lasts for $d_2 = 4$ time units in this instance, which produces four observations (o_3, o_4, o_5, o_6) according to the emission probability $b_{i_2,d_2}(o_3, o_4, o_5, o_6)$.
5. (i_2, d_2) then transits to (i_3, d_3), and (i_3, d_3) transits to ..., (i_N, d_N) until the final observation o_T is produced. The last state i_N lasts for d_N time units, where $\sum_{n=1}^{N} d_n = T$ and T is the total number of observations.

Note that the underlying stochastic process is not observable, that is, the states and their durations are hidden and there is no external demarcation between the observations arising from state i_1 and those arising from state i_2. Only the sequence of observations $(o_1, o_2, \ldots, o_T)$ can be observed.

Example 1.4

The observed workload data of a Web site consists of o_t, the number of user requests per second, with the maximum observed value $K = \max\{o_t\} = 74$ requests/s. The total number of observations is $T = 3600$ s (over 1 h) in the workload data set. The user request arrivals are governed by an underlying hidden semi-Markov process. The hidden state represents the arrival rate, which is corresponding to the number of users that are browsing the website. For instance, state 1 corresponds to arrival rate of 13 requests/s, state10, 30.1 requests/s, and state 20, 60

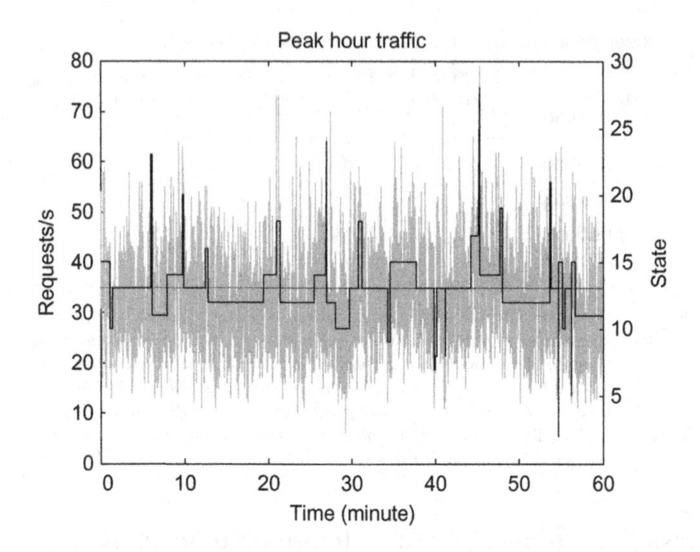

Figure 1.7 Data (requests/s) and the hidden states of the workload.
The grey line is the trace of the workload data (number of arrivals per second) that we observed. The black line is the hidden state sequence estimated. The hidden state represents the arrival rate, and the state duration represents the dwell time of the arrival rate. For given arrival rate, the number of arrivals per second is a random variable.

requests/s. For given state or arrival rate, the number of user requests per second, O_t, is a random variable. Therefore, from an observed value one cannot tell the actual arrival rate. In other words, the state is hidden. Figure 1.7 plots the observed number of requests per second together with the estimated hidden states (i.e., arrival rates). There are 41 state transitions occurred during the period of 3600 s. Some states last for a long period of time with the maximum duration of $D = 405$ s (Yu et al., 2002).

The issues related to a general HSMM include:

1. computation of the predicted/filtered/smoothed probabilities, expectations, and the likelihood of observations;
2. the MAP estimation of states and the maximum likely state sequence estimation; and
3. parameter estimation/update.

Different models have different capability in modeling applications and have different computational complexity in solving those issues.

1.6 HISTORY OF HIDDEN SEMI-MARKOV MODELS

The first approach to HSMM was proposed by Ferguson (1980), which is partially included in the survey paper by Rabiner (1989).

This approach is called the explicit duration HMM in contrast to the implicit duration of the HMM. It assumes that the state duration is generally distributed depending on the current state of the underlying semi-Markov process. It also assumes the "conditional independence" of outputs. Levinson (1986a) replaced the probability mass functions of duration with continuous probability density functions to form a continuously variable duration HMM. As Ferguson (1980) pointed out, an HSMM can be realized in the HMM framework in which both the state and its sojourn time since entering the state are taken as a complex HMM state. This idea was exploited in 1991 by a 2-vector HMM (Krishnamurthy et al., 1991) and a duration-dependent state transition model (Vaseghi, 1991). Since then, similar approaches were proposed in many applications. They are called in different names such as inhomogeneous HMM (Ramesh and Wilpon, 1992), nonstationary HMM (Sin and Kim, 1995), and triplet Markov chains (Pieczynski et al., 2002). These approaches, however, have the common problem of computational complexity in some applications. A more efficient algorithm was proposed in 2003 by Yu and Kobayashi (2003a), in which the forward–backward variables are defined using the notion of a state together with its remaining sojourn (or residual life) time. This makes the algorithm practical in many applications.

The HSMM has been successfully applied in many areas. The most successful application is in speech recognition. The first application of HSMM in this area was made by Ferguson (1980). Since then, there have been more than one hundred such papers published in the literature. It is the application of HSMM in speech recognition that enriches the theory of HSMM and develops many algorithms for HSMM.

Since the beginning of 1990s, the HSMM started being applied in many other areas. In this decade, the main application area of HSMMs is handwritten/printed text recognition (see, e.g., Chen et al., 1993a). Other application areas of HSMMs include electrocardiograph (ECG) (Thoraval et al., 1992), network traffic characterization (Leland et al., 1994), recognition of human genes in DNA (Kulp et al., 1996), language identification (Marcheret and Savic, 1997), ground target tracking (Ke and Llinas, 1999), document image comparison, and classification at the spatial layout level (Hu et al., 1999).

From 2000 to 2009, the HSMM has been obtained more and more attentions from vast application areas. In this decade, the main applications are human activity recognition (see, e.g., Hongeng and Nevatia, 2003) and speech synthesis (see, e.g., Moore and Savic, 2004). Other application areas include change-point/end-point detection for semiconductor manufacturing (Ge and Smyth, 2000a), protein structure prediction (Schmidler et al., 2000), analysis of branching and flowering patterns in plants (Guedon et al., 2001), rain events time series model (Sansom and Thomson, 2001), brain functional MRI sequence analysis (Faisan et al., 2002), Internet traffic modelling (Yu et al., 2002), event recognition in videos (Hongeng and Nevatia, 2003), image segmentation (Lanchantin and Pieczynski, 2004), semantic learning for a mobile robot (Squire, 2004), anomaly detection for network security (Yu, 2005), symbolic plan recognition (Duong et al., 2005a), terrain modeling (Wellington et al., 2005), adaptive cumulative sum test for change detection in noninvasive mean blood pressure trend (Yang et al., 2006), equipment prognosis (Bechhoefer et al., 2006), financial time series modeling (Bulla and Bulla, 2006), remote sensing (Pieczynski, 2007), classification of music (Liu et al., 2008), and prediction of particulate matter in the air (Dong et al., 2009).

In the recent years since 2010, the main application areas of HSMMs are equipment prognosis/diagnosis (see, e.g., Dong and Peng, 2011) and animal activity modeling (see, e.g., O'Connell et al., 2011). Other application areas include such as machine translation (Bansal et al., 2011), network performance (Wang et al., 2011), deep brain stimulation (Taghva, 2011), image recognition (Takahashi et al., 2010), icing load prognosis (Wu et al., 2014), irrigation behavior (Andriyas and McKee, 2014), dynamics of geyser (Langrock, 2012), anomaly detection of spacecraft (Tagawa et al., 2011), and prediction of earthquake (Beyreuther and Wassermann, 2011).

CHAPTER 2

General Hidden Semi-Markov Model

This chapter provides a unified description of hidden semi-Markov models, and discusses important issues related to inference in the HSMM.

2.1 A GENERAL DEFINITION OF HSMM

An HSMM allows the underlying process to be a semi-Markov chain with a variable *duration* or *sojourn time* for each state. State duration d is a random variable and assumes an integer value in the set $\mathbf{D} = \{1, 2, \ldots, D\}$, where D is the maximum duration of a state and can be infinite in some applications. Each state can emit a series of observations, and the number of observations produced while in state i is determined by the length of time spent in the state, that is, the duration d. Now we provide a unified description of HSMMs.

Assume a discrete-time semi-Markov process with a set of (hidden) states $\mathbf{S} = \{1, \ldots, M\}$. The state sequence $(S_1, \ldots, S_T)$ is denoted by $S_{1:T}$, where $S_t \in \mathbf{S}$ is the state at time t. A realization of $S_{1:T}$ is denoted as $s_{1:T}$. For simplicity of notation in the following sections, we denote:

- $S_{t_1:t_2} = i$—state i that the system stays in during the period from t_1 to t_2. In other words, it means $S_{t_1} = i, S_{t_1+1} = i, \ldots,$ and $S_{t_2} = i$. Note that the previous state S_{t_1-1} and the next state S_{t_2+1} may or may not be i.
- $S_{[t_1:t_2]} = i$—state i that starts at time t_1 and ends at t_2 with duration $d = t_2 - t_1 + 1$. This implies that the previous state S_{t_1-1} and the next state S_{t_2+1} must not be i.
- $S_{[t_1:t_2} = i$—state i that starts at time t_1 and lasts till t_2, with $S_{[t_1} = i, S_{t_1+1} = i, \ldots, S_{t_2} = i$, where $S_{[t_1} = i$ means that at t_1 the system entered state i from some other state, that is, the previous state S_{t_1-1} must not be i. The next state S_{t_2+1} may or may not be i.
- $S_{t_1:t_2]} = i$—state i that lasts from t_1 to t_2 and ends at t_2 with $S_{t_1} = i, S_{t_1+1} = i, \ldots, S_{t_2]} = i$, where $S_{t_2]} = i$ means that at time t_2 the state will end and transit to some other state at time $t_2 + dt_2$, that is, the next state S_{t_2+1} must not be i. The previous state S_{t_1-1} may or may not be i.

Based on these definitions, $S_{[t]} = i$ means state i starting and ending at t with duration 1, $S_{[t} = i$ means state i starting at t, $S_{t]} = i$ means state i ending at t, and $S_t = i$ means the state at t being state i.

Denote the observation sequence $(O_1, \ldots, O_T)$ by $O_{1:T}$, where $O_t \in \mathbf{V}$ is the observation at time t and $\mathbf{V} = \{v_1, v_2, \ldots, v_K\}$ is the set of observable values. For observation sequence $O_{1:T}$, the underlying state sequence is $S_{1:d_1]} = i_1$, $S_{[d_1+1:d_1+d_2]} = i_2, \ldots, S_{[d_1+\cdots+d_{N-1}+1:d_1+\cdots+d_N} = i_N$, and the state transitions are $(i_n, d_n) \to (i_{n+1}, d_{n+1})$, for $n = 1, \ldots, N-1$, where $\sum_{n=1}^{N} d_n = T$, $i_1, \ldots, i_N \in \mathbf{S}$, and $d_1, \ldots, d_N \in \mathbf{D}$. Note that the first state i_1 is not necessarily starting at time 1 associated with the first observation O_1 and the last state i_N is not necessarily ending at time T associated with the last observation O_T. As the states are hidden, the number N of hidden states in the underlying state sequence is also hidden/unknown.

We note that the observable values can be discrete, continuous, or have infinite support, and the observation $O_t \in \mathbf{V}$ can be a value, a vector, a symbol, or an event. The length T of the observation sequence can be very large, but is usually assumed to be finite except in the case of online learning. There are usually multiple observation sequences in practice, but we do not always explicitly mention this fact unless it is required. The formulas derived for the single observation sequence usually cannot be directly applied for the multiple observation sequences because the sequence lengths are different with different likelihood functions. Therefore, while applying the formulas derived for the single observation sequence into the case of multiple observation sequences, the formulas must be divided by the likelihood functions $P[o_{1:T_l}^{(l)}]$ if they have not yet appeared in the formulas, where $o_{1:T_l}^{(l)}$ is the lth observation sequence of length T_l.

Suppose the current time is t, the process has made $n-1$ jumps, and the time spent since the previous jump is $X_t = t - \sum_{l=1}^{n-1} d_l$. As explained in Section 1.1.2, the process $(S_t, X_t)_{t \geq 1}$ is a discrete-time homogeneous Markov process. Its subsequence $(i_n, d_n)_{n \geq 1}$ is also a Markov process based on the Markov property. Then we can define the state transition probability from state i having duration h to state $j \neq i$ having duration d by

$$a_{(i,h)(j,d)} \equiv P[S_{[t+1:t+d]} = j | S_{[t-h+1:t]} = i],$$

which is assumed independent of time t, for $i, j \in \mathbf{S}$, $h, d \in \mathbf{D}$. The transition probabilities must satisfy $\sum_{j \in \mathbf{S} \setminus \{i\}} \sum_{d \in \mathbf{D}} a_{(i,h)(j,d)} = 1$, for all given $i \in \mathbf{S}$ and $h \in \mathbf{D}$, with zero self-transition probabilities $a_{(i,h)(i,d)} = 0$, for all $i \in \mathbf{S}$ and $h, d \in \mathbf{D}$. In other words, when a state ends at time t, it cannot transit to the same state at the next time $t+1$ because the state durations are explicitly specified by some distributions other than geometric or exponential distributions. From the definition we can see that the previous state i started at $t-h+1$ and ended at t, with duration h. Then it transits to state j having duration d, according to the state transition probability $a_{(i,h)(j,d)}$. State j will start at $t+1$ and end at $t+d$. This means both the state and the duration are dependent on both the previous state and its duration. While in state j, there will be d observations $O_{t+1:t+d}$ being emitted. Denote this emission/observation probability by

$$b_{j,d}(o_{t+1:t+d}) \equiv P[o_{t+1:t+d} | S_{[t+1:t+d]} = j]$$

which is assumed to be independent of time t, where $o_{t+1:t+d}$ is the observed values of $O_{t+1:t+d}$. Let the distribution of the first state be

$$\Pi_{j,d} \equiv P[S_{[1:d]} = j]$$

or

$$\Pi_{j,d} \equiv P[S_{1:d]} = j]$$

depending on the model assumption that the first state is starting at $t=1$ or before. We can equivalently let the initial distribution of the state be

$$\pi_{j,d} \equiv P[S_{[t-d+1:t]} = j], \quad t \leq 0.$$

It represents the probability of the initial state and its duration before time $t=1$ or before the first observation o_1 obtained. The relationship between the two definitions of initial state distribution is $\Pi_{j,d} = \sum_{\tau=d-D+1}^{1} \sum_{i,h} \pi_{(i,h)} a_{(i,h)(j,d-\tau+1)}$, where if the starting time of the first state must be $t=1$ then $\tau = 1$; otherwise, if the first state can start at or before $t=1$, then $1 \geq \tau \geq -(D-d-1)$. Usually, the second definition of the initial state distribution, $\{\pi_{j,d}\}$, makes the computation of the forward variables in the HSMM algorithms simpler. Then the set of model parameters for the HSMM is defined by

$$\lambda \equiv \{a_{(i,h)(j,d)}, b_{j,d}(v_{k_1:k_d}), \pi_{j,d} : i,j \in \mathbf{S}; h,d \in \mathbf{D}; v_{k_d} \in \mathbf{V}\},$$

or

$$\lambda \equiv \{a_{(i,h)(j,d)}, b_{j,d}(v_{k_1:k_d}), \Pi_{j,d} : i,j \in \mathbf{S}; h,d \in \mathbf{D}; v_{k_d} \in \mathbf{V}\},$$

where $v_{k_1:k_d}$ represents an observable substring of length d for $v_{k_1}\ldots v_{k_d} \in \mathbf{V}^d = \mathbf{V} \times \ldots \times \mathbf{V}$. This general HSMM is shown in Figure 1.6.

The general HSMM is reduced to specific models of HSMM depending on the assumptions made. For instances,

1. If the state duration is assumed to be independent of the previous state, then the state transition probability can be further specified as $a_{(i,h)(j,d)} = a_{(i,h)j}p_j(d)$, where

$$a_{(i,h)j} \equiv P[S_{[t+1} = j | S_{[t-h+1:t]} = i] \tag{2.1}$$

is the transition probability from state i that has stayed for duration h to state j that will start at $t+1$, and

$$p_j(d) \equiv P[S_{t+1:t+d]} = j | S_{[t+1} = j] \tag{2.2}$$

is the probability of duration d that state j will take. This is the model proposed by Marhasev et al. (2006). Compared with the general HSMM, the number of model parameters is reduced from M^2D^2 to $M^2D + MD$, and the state duration distributions $p_j(d)$ can be explicitly expressed using probability density functions (e.g., Gaussian distributions) or a probability mass function.
2. If a state transition is assumed to be independent of the duration of the previous state, then the state transition probability from (i,h) to (j,d) becomes $a_{(i,h)(j,d)} = a_{i(j,d)}$, where

$$a_{i(j,d)} \equiv P[S_{[t+1:t+d]} = j | S_{t]} = i] \tag{2.3}$$

is the transition probability that state i ended at t and transits to state j having duration d. If it is assumed that a state transition for $i \neq j$ is $(i,1) \rightarrow (j,d)$ and a self-transition is $(i,d+1) \rightarrow (i,d)$, for $d \in \mathbf{D}$, then the model becomes the residual time HMM (Yu and Kobayashi, 2003a). In this model, the starting time of the state is not of concern, but the ending time is of interest. Therefore, d represents the remaining sojourn (or residual life) time of state j. This model is obviously appropriate to applications for which the residual life is of the most concern. The number of model parameters is reduced to M^2D. More importantly, if the state duration is further assumed to be independent of the previous state, then the state transition probability can be specified as $a_{i(j,d)} = a_{i,j}p_j(d)$. In this case, the computational complexity will be the lowest among all HSMMs. The number of model parameters is further reduced to $M^2 + MD$.

3. If self-transition $(i,d) \rightarrow (i,d+1)$ is allowed and the state duration is assumed to be independent of the previous state, then the state transition probability becomes

$$a_{(i,h)(j,d)} = a_{(i,h)j}\left(\prod_{\tau=1}^{d-1} a_{jj}(\tau)\right)[1 - a_{jj}(d)],$$

where $a_{(i,h)j} \equiv P[S_{[t+1} = j | S_{[t-h+1:t]} = i]$; $a_{jj}(d)$ is the self-transition probability when state j has stayed for d time units, that is,

$$a_{jj}(d) \equiv P[S_{t+d+1} = j | S_{[t+1:t+d} = j],$$

and $1 - a_{jj}(d) = P[S_{t+d]} = j | S_{[t+1:t+d} = j]$ is the probability that state j ends with duration d. This is the variable transition HMM (Krishnamurthy et al., 1991; Vaseghi, 1991). In this model, a state transition is either $(i,d) \rightarrow (j,1)$ for $i \neq j$ or $(i,d) \rightarrow (i,d+1)$ for a self-transition. This process is similar to the standard discrete-time semi-Markov process. The concept of the discrete-time semi-Markov process can thus be used in modeling an application. This model has $M^2D + MD$ model parameters. The computational complexity is relatively high compared with other conventional HSMMs.

4. If a transition to the current state is independent of the duration of the previous state and the duration of the current state is only conditioned on the current state itself, then

$$a_{(i,h)(j,d)} = a_{ij}p_j(d),$$

where $a_{ij} \equiv P[S_{[t+1} = j | S_{t]} = i]$ is the transition probability from state i to state j, with the self-transition probability $a_{ii} = 0$. This is the explicit duration HMM (Ferguson, 1980), with $M^2 + MD$ model parameters and lower computational complexity. This is the simplest and the most popular model among all HSMMs, with easily understandable formulas and modeling concepts.

Besides, the general form $b_{j,d}(v_{k_1:k_d})$ of observation distributions can be simplified and dedicated to applications. They can be parametric (e.g., a mixture of Gaussian distributions) or nonparametric (e.g., a probability mass function), discrete or continuous, and dependent on or independent of the state durations. The observations can be assumed dependent or conditionally independent for given states, that is, $b_{j,d}(v_{k_1:k_d}) = \prod_{\tau=1}^{d} b_j(v_{k_\tau})$. The conditional independence makes HSMMs simpler and so is often assumed in the literature.

2.2 FORWARD–BACKWARD ALGORITHM FOR HSMM

For an observation sequence $o_{1:T}$, the likelihood function for given model parameters λ is

$$P[o_{1:T}|\lambda] = \sum_{S_{1:T}} P[S_{1:T}, o_{1:T}|\lambda].$$

Suppose $S_{1:T} = (i_1, d_1)\cdots(i_N, d_N)$, satisfying $\sum_{n=1}^{N} d_n = T$. Let $t_n = \sum_{m=1}^{n} d_m$. Then $P[S_{1:T}, o_{1:T}|\lambda] = \prod_{n=1}^{N} a_{(i_{n-1},d_{n-1})(i_n,d_n)} b_{i_n,d_n}(o_{t_{n-1}+1:t_n})$, where $a_{(i_0,d_0)(i_1,d_1)} = \Pi_{i_1,d_1}$ for simplicity. To sum over all possible $S_{1:T}$, for all $N \geq 1$, $i_1, \ldots, i_N \in \mathbf{S}$ and $d_1, \ldots, d_N \in \mathbf{D}$, the computational amount involved will be huge. Therefore, a sum-product form algorithm, that is, a forward–backward algorithm is usually used in the literature. Now we define the forward and backward variables.

The forward variables for HSMM are defined by

$$\alpha_t(j,d) \equiv P[S_{[t-d+1:t]} = j, o_{1:t}|\lambda] \tag{2.4}$$

and the backward variables by

$$\beta_t(j,d) \equiv P[o_{t+1:T}|S_{[t-d+1:t]} = j, \lambda]. \tag{2.5}$$

Based on the Markov property, the current/future observations are dependent on the current state, for example,

$$P[o_{t-d+1:t}|S_{[t-d-h+1:t-d]} = i, S_{[t-d+1:t]} = j, \lambda] = P[o_{t-d+1:t}|S_{[t-d+1:t]} = j, \lambda]$$

and

$$P[o_{t+1:T}|S_{[t-d+1:t]} = j, S_{[t+1:t+h]} = i, \lambda] = P[o_{t+1:T}|S_{[t+1:t+h]} = i, \lambda],$$

and independent of the previous observations, for example,

$$P[S_{[t-d+1:t]} = j, o_{t-d+1:t} | S_{[t-d-h+1:t-d]} = i, o_{1:t-d}, \lambda]$$
$$= P[S_{[t-d+1:t]} = j, o_{t-d+1:t} | S_{[t-d-h+1:t-d]} = i, \lambda]$$

and

$$P[o_{t+h+1:T} | S_{[t+1:t+h]} = i, o_{t+1:t+h}, \lambda] = P[o_{t+h+1:T} | S_{[t+1:t+h]} = i, \lambda].$$

Using these equations, it is easy to obtain the forward–backward algorithm for a general HSMM:

$$\begin{aligned} \alpha_t(j,d) &= \sum_{i \neq j; h} P[S_{[t-d-h+1:t-d]} = i, S_{[t-d+1:t]} = j, o_{1:t} | \lambda] \\ &= \sum_{i \neq j; h} \alpha_{t-d}(i,h) \cdot P[S_{[t-d+1:t]} = j, o_{t-d+1:t} | S_{[t-d-h+1:t-d]} = i, \lambda] \\ &= \sum_{i \neq j; h} \alpha_{t-d}(i,h) \cdot a_{(i,h)(j,d)} \cdot P[o_{t-d+1:t} | S_{[t-d+1:t]} = j, \lambda] \\ &= \sum_{i \neq j; h} \alpha_{t-d}(i,h) \cdot a_{(i,h)(j,d)} \cdot b_{j,d}(o_{t-d+1:t}), \end{aligned} \tag{2.6}$$

for $t > 0$, $d \in \mathbf{D}$, $j \in \mathbf{S}$, and

$$\begin{aligned} \beta_t(j,d) &= \sum_{i \neq j; h} P[S_{[t+1:t+h]} = i, o_{t+1:T} | S_{[t-d+1:t]} = j, \lambda] \\ &= \sum_{i \neq j; h} a_{(j,d)(i,h)} \cdot P[o_{t+1:T} | S_{[t+1:t+h]} = i, \lambda] \\ &= \sum_{i \neq j; h} a_{(j,d)(i,h)} \cdot b_{i,h}(o_{t+1:t+h}) \cdot P[o_{t+h+1:T} | S_{[t+1:t+h]} = i, \lambda] \\ &= \sum_{i \neq j; h} a_{(j,d)(i,h)} \cdot b_{i,h}(o_{t+1:t+h}) \cdot \beta_{t+h}(i,h), \end{aligned} \tag{2.7}$$

for $t < T$, $d \in \mathbf{D}$, $j \in \mathbf{S}$.

2.2.1 Symmetric Form of the Forward–Backward Algorithm

Though the backward formula (2.7) seems a little bit different from the forward formula (2.6), it can be transformed to the same form as the forward one. Now we derive the symmetric form of the forward–backward algorithm. If we use the starting time of the given state to express the backward variables, that is, let $\beta''_{t-d+1}(j,d) = b_{j,d}(o_{t-d+1:t})\beta_t(j,d)$, then the backward formula (2.7) becomes

$$\beta''_{t-d+1}(j,d) = b_{j,d}(o_{t-d+1:t}) \sum_{i \neq j;h} a_{(j,d)(i,h)} \beta''_{t+1}(i,h)$$

or

$$\beta''_t(j,d) = b_{j,d}(o_{t:t+d-1}) \sum_{i \neq j;h} a_{(j,d)(i,h)} \beta''_{t+d}(i,h).$$

If we further denote the backward variables in the reverse time order, that is, let $t' = T - t + 1$, $\beta'_t(j,d) = \beta''_{T-t+1}(j,d)$, $o'_t = o_{T-t+1}$, and $a'_{(i,h)(j,d)} = a_{(j,d)(i,h)}$, then the backward formula becomes

$$\begin{aligned} \beta'_t(j,d) &= b_{j,d}(o_{T-t+1:T-t+1+d-1}) \sum_{i \neq j;h} a_{(j,d)(i,h)} \beta''_{T-t+1+d}(i,h) \\ &= b_{j,d}(o'_{t:t-d+1}) \sum_{i \neq j;h} a_{(j,d)(i,h)} \beta'_{t-d}(i,h) \\ &= \sum_{i \neq j;h} \beta'_{t-d}(i,h) a'_{(i,h)(j,d)} b_{j,d}(o'_{t-d+1:t}). \end{aligned}$$

We can see that the backward recursion is exactly the same as the forward formula (2.6) when it is expressed in the reverse time order. This can potentially reduce the requirement for the silicon area on a chip if the backward logic module uses the forward one. A symmetric forward–backward algorithm for the residual time model was introduced by Yu and Kobayashi (2003a).

2.2.2 Initial Conditions

The initial conditions generally can have two different assumptions:

1. *The general assumption of boundary conditions* assumes that the first state begins at or before observation o_1 and the last state ends at or after observation o_T. In this case, we can assume that the process starts at $-\infty$ and terminates at $+\infty$. The observations out of the sampling period [1,T] can be any possible values, that is, $b_{j,d}(\cdot) = 1$

for any $j \in \mathbf{S}, d \in \mathbf{D}$. Therefore, in the forward formula (2.6) $b_{j,d}(o_{t-d+1:t})$ is replaced with the distribution $b_{j,d}(o_{1:t})$ if $t-d+1 \leq 1$ and $t \geq 1$, and in the backward formula (2.7) $b_{i,h}(o_{t+1:t+h})$ is replaced with $b_{i,h}(o_{t+1:T})$ if $t+1 \leq T$ and $t+h \geq T$. We then have the initial conditions for the forward recursion formula (2.6) as follows:

$$\alpha_\tau(j,d) = P[S_{[\tau-d+1:\tau]} = j|\lambda] = \pi_{j,d}, \quad \tau \leq 0, d \in \mathbf{D}, \tag{2.8}$$

where $\{\pi_{j,d}\}$ can be the equilibrium distribution of the underlying semi-Markov process. Because, for $t+h \geq T$,

$$P[S_{[t+1:t+h]} = i, o_{t+1:T}|S_{[t-d+1:t]} = j, \lambda] = a_{(j,d)(i,h)} b_{i,h}(o_{t+1}^T),$$

then from the backward recursion formula (2.7) we can see that $\beta_{t+h}(i,h) = 1$, for $t+h \geq T$. Therefore, the initial conditions for the backward recursion formula (2.7) are as follows:

$$\beta_\tau(i,d) = 1, \quad \tau \geq T, d \in \mathbf{D}. \tag{2.9}$$

If the model assumes that the first state begins at $t=1$ and the last state ends at or after observation o_T, it is a right-censored HSMM introduced by Guedon (2003). Because this is desirable for many applications, it is taken as a basis for an R package for analyzing HSMMs (Bulla et al., 2010).

2. *The simplified assumption of boundary conditions* assumes that the first state begins at time 1 and the last state ends at time T. This is the most popular assumption one can find in the literature. In this case, the initial conditions for the forward recursion formula (2.6) are

$$\begin{aligned} &\alpha_0(j,d) = \pi_{j,d}, && d \in \mathbf{D}, \\ &\alpha_\tau(j,d) = 0, && \tau < 0, d \in \mathbf{D}, \end{aligned}$$

and the initial conditions for the backward recursion formula (2.7) are

$$\begin{aligned} &\beta_T(i,d) = 1, && d \in \mathbf{D}, \\ &\beta_\tau(i,d) = 0, && \tau > T, d \in \mathbf{D}. \end{aligned}$$

Note that the initial distributions of states can be assumed as $\Pi_{j,d} \equiv P[S_{[1:d]} = j|\lambda]$. Therefore, the initial conditions for the forward recursion formula can be changed to $\alpha_d(j,d) = \Pi_{j,d} b_{j,d}(o_{1:d})$, for $d \in \mathbf{D}$, and all others $\alpha_t(j,d)$, for $t \neq d$ and $t \leq D$, being zeros.

Therefore, the forward–backward algorithm for the general HSMM is as follows, where the self-transition probabilities a(i;h)(i;d) = 0, for all *i*, *h*, and *d*:

Algorithm 2.1 Forward–Backward Algorithm for the General HSMM

The Forward Algorithm

1. *For* $j = 1, \ldots, M$ *and* $d = 1, \ldots, D$, *let* $\alpha_0(j,d) = \pi_{j,d}$;
2. *If the simplified assumption that the first state must start at* $t = 1$ *is assumed, let* $\alpha_\tau(j,d) = 0$ *for* $\tau < 0$; *otherwise let* $\alpha_\tau(j,d) = \pi_{j,d}$ *for* $\tau < 0$;
3. *For* $t = 1, \ldots, T$ {
   for $j = 1, \ldots, M$ *and* $d = 1, \ldots, D$ {

$$\alpha_t(j,d) = \sum_{i,h} a_{t-d}(i,h) a_{(i,h)(j,d)} b_{j,d}(o_{t-d+1:t});$$

   }
}

The Backward Algorithm

1. *For* $j = 1, \ldots, M$ *and* $d = 1, \ldots, D$, *let* $\beta_T(j,d) = 1$;
2. *If the simplified assumption that the last state must end at* $t = T$ *is assumed, let* $\beta_\tau(j,d) = 0$ *for* $\tau > T$; *otherwise, let* $\beta_\tau(j,d) = 1$ *for* $\tau > T$.
3. *For* $t = T - 1, \ldots, 1$ {
   for $j = 1, \ldots, M$ *and* $d = 1, \ldots, D$ {

$$\beta_t(j,d) = \sum_{i,h} a_{(j,d)(i,h)} b_{i,h}(o_{t+1:t+d}) \beta_{t+d}(i,h);$$

   }
}

2.2.3 Probabilities

After the forward variables $\{\alpha_t(j,d)\}$ and the backward variables $\{\beta_t(j,d)\}$ are determined, all other probabilities of interest can be computed. For instances, the filtered probability that state j starts at $t-d+1$ and ends at t, with duration d, given partial observed sequence $o_{1:t}$, can be determined by

$$P\left[S_{[t-d+1:t]} = j | o_{1:t}, \lambda\right] = \frac{\alpha_t(j,d)}{P[o_{1:t}|\lambda]},$$

and the predicted probability that state j will start at $t+1$ and end at $t+d$, with duration d, given partial observed sequence $o_{1:t}$ by

$$P\left[S_{[t+1:t+d]}=j|o_{1:t},\lambda\right]=\frac{\sum_{i\neq j,h}\alpha_t(i,h)a_{(i,h)(j,d)}}{P[o_{1:t}|\lambda]},$$

where

$$\begin{aligned}P[o_{1:t}|\lambda]&=\sum_j P[S_t=j,o_{1:t}|\lambda]=\sum_j\sum_d\sum_{0\leq k\leq D-d}P[S_{[t-d+1:t+k]}=j,o_{1:t}|\lambda]\\&=\sum_j\sum_d\sum_{0\leq k\leq D-d}\sum_{o_{t+1:t+k}}P[S_{[t-d+1:t+k]}=j,o_{1:t+k}|\lambda]\\&=\sum_j\sum_d\sum_{i\neq j;h}\alpha_{t-d}(i,h)\sum_{0\leq k\leq D-d}a_{(i,h)(j,d+k)}\sum_{o_{t+1:t+k}}b_{j,d+k}(o_{t-d+1:t+k}).\end{aligned}$$

These readily yield the filtered probability of state j ending at t, $P[S_{t]}=j|o_{1:t},\lambda]=\sum_d P[S_{[t-d+1:t]}=j|o_{1:t},\lambda]$, and the predicted probability of state j starting at $t+1$, $P[S_{[t+1}=j|o_{1:t},\lambda]=\sum_d P[S_{[t+1:t+d]}=j|o_{1:t},\lambda]$.

The smoothed or posterior probabilities, such as $P[S_t=j|o_{1:T},\lambda]$, $P[S_t=i,S_{t+1}=j|o_{1:T},\lambda]$ and $P[S_{[t-d+1:t]}=j|o_{1:T},\lambda]$, for given entire observation sequence $o_{1:T}$ and model parameters λ can be determined by the following equations, for $h,d\in\mathbf{D}$, $i,j\in\mathbf{S}$, $i\neq j$, and $t=1,\ldots,T$,

$$\eta_t(j,d)\equiv P[S_{[t-d+1:t]}=j,o_{1:T}|\lambda]=\alpha_t(j,d)\beta_t(j,d),\qquad(2.10)$$

representing the joint probability that the observation sequence is $o_{1:T}$ and the state is j having duration d by time t given the model,

$$\begin{aligned}\xi_t(i,h;j,d)&\equiv P[S_{[t-h+1:t]}=i,S_{[t+1:t+d]}=j,o_{1:T}|\lambda]\\&=\alpha_t(i,h)a_{(i,h)(j,d)}b_{j,d}(o_{t+1:t+d})\beta_{t+d}(j,d),\end{aligned}\qquad(2.11)$$

representing the joint probability that the observation sequence is $o_{1:T}$ and the transition from state i of duration h to state j of duration d occurs at time t given the model,

$$\begin{aligned}\xi_t(i,j)&\equiv P[S_{t]}=i,S_{[t+1}=j,o_{1:T}|\lambda]\\&=\sum_{h\in D}\sum_{d\in D}\xi_t(i,h;j,d),\end{aligned}\qquad(2.12)$$

representing the joint probability that the observation sequence is $o_{1:T}$ and the transition from state i to state j occurs at time t given the model,

$$\gamma_t(j) \equiv P[S_t = j, o_{1:T}|\lambda] = \sum_{\substack{\tau, d: \\ \tau \geq t \geq \tau - d + 1}} P[S_{[\tau-d+1:\tau]} = j, o_{1:T}|\lambda] = \sum_{\tau \geq t} \sum_{d=\tau-t+1}^{D} \eta_\tau(j, d), \tag{2.13}$$

representing the joint probability that the observation sequence is $o_{1:T}$ and the state is j at time t given the model, and

$$P[o_{1:T}|\lambda] = \sum_{j \in S} P[S_t = j, o_{1:T}|\lambda] = \sum_{j \in S} \gamma_t(j),$$

being the likelihood probability that the observed sequence $o_{1:T}$ is generated by the model λ. Then, the smoothed probabilities can be obtained by letting:

$\overline{\eta}_t(j, d) = \eta_t(j, d)/P[o_{1:T}|\lambda]$ be the smoothed probability of being in state j having duration d by time t given the model and the observation sequence;
$\overline{\xi}_t(i, h; j, d) = \xi_t(i, h; j, d)/P[o_{1:T}|\lambda]$ the smoothed probability of transition at time t from state i occurred with duration h to state j having duration d given the model and the observation sequence;
$\overline{\xi}_t(i, j) = \xi_t(i, j)/P[o_{1:T}|\lambda]$ the smoothed probability of transition at time t from state i to state j given the model and the observation sequence; and
$\overline{\gamma}_t(j) = \gamma_t(j)/P[o_{1:T}|\lambda]$ the smoothed probability of being in state j at time t given the model and the observation sequence.

Obviously, the conditional factor $P[o_{1:T}|\lambda]$ is common for all the smoothed/posterior probabilities. Therefore, it is often omitted for simplicity in the literature. Similarly, in the rest of this book, this conditional factor is sometimes not explicitly mentioned while calculating the smoothed/posterior probabilities.

In considering the following identity

$$P[S_{t:t+1} = j, o_{1:T}|\lambda] = P[S_t = j, o_{1:T}|\lambda] - P[S_{t]} = j, o_{1:T}|\lambda]$$

$$P[S_{t:t+1} = j, o_{1:T}|\lambda] = P[S_{t+1} = j, o_{1:T}|\lambda] - P[S_{[t+1} = j, o_{1:T}|\lambda],$$

we have a recursive formula for calculating $\gamma_t(j)$:

$$\begin{aligned}\gamma_t(j) &= \gamma_{t+1}(j) + P[S_{t]} = j, o_{1:T}|\lambda] - P[S_{[t+1} = j, o_{1:T}|\lambda] \\ &= \gamma_{t+1}(j) + \sum_{i \in \mathbf{S}\setminus\{j\}} \left[\xi_t(j,i) - \xi_t(i,j)\right].\end{aligned} \tag{2.14}$$

2.2.4 Expectations

Using the forward and backward variables, one can compute various conditional expectations given $o_{1:T}$ (Ferguson, 1980):

1. The expected number of times that state i ends at or before t: $\sum_{\tau \le t} \sum_{j \in \mathbf{S}\setminus\{i\}} \overline{\xi}_\tau(i,j)$; The expected number of times that state i starts at or before t: $\sum_{\tau \le t-1} \sum_{j \in \mathbf{S}\setminus\{i\}} \overline{\xi}_\tau(j,i)$;
2. The expected total number of times that state i commenced, $\sum_t \sum_{j \in \mathbf{S}\setminus\{i\}} \overline{\xi}_t(j,i)$, or terminated, $\sum_t \sum_{j \in \mathbf{S}\setminus\{i\}} \overline{\xi}_t(i,j)$. For the simplified assumption of the boundary conditions, $\sum_{t=0}^{T-1} \sum_{j \in \mathbf{S}\setminus\{i\}} \overline{\xi}_t(j,i) = \sum_{t=1}^{T} \sum_{j \in \mathbf{S}\setminus\{i\}} \overline{\xi}_t(i,j)$;
3. The expected total duration spent in state i: $\sum_t \overline{\gamma}_t(i)$;
4. The expected average duration of state i: $\frac{\sum_t \sum_d \eta_t(i,d) d}{\sum_t \sum_d \eta_t(i,d)}$;
5. The expected number of times that state i occurred with observation $o_t = v_k$: $\sum_t \overline{\gamma}_t(i) I(o_t = v_k)$, where the indicator function $I(x) = 1$ if x is true and 0 otherwise;
6. The expected average observable value of state i: $\frac{\sum_t \gamma_t(i) o_t}{\sum_t \gamma_t(i)}$;
7. The smoothed probability that state i was the first state: $\overline{\gamma}_1(i)$.

2.2.5 MAP Estimation of States

The MAP estimation of state S_t given a specific observation sequence $o_{1:T}$ can be obtained (Ferguson, 1980) by finding the maximum $\gamma_t(j)$ given by Eqn (2.13) or (2.14), that is,

$$\hat{s}_t = \arg\max_{i \in \mathbf{S}} \{\bar{\gamma}_t(i)\} = \arg\max_{i \in \mathbf{S}} \{\gamma_t(i)\}. \tag{2.15}$$

To find out in practice when the errors are most likely to occur in the state estimation, the confidence of the state estimation can be simply defined as

$$(\gamma_t(\hat{s}_t) - \max_{i \neq \hat{s}_t} \{\gamma_t(i)\}) / \gamma_t(\hat{s}_t).$$

By using Eqn (2.15) for $t = 1, \ldots, T$, the state sequence can be estimated as $\hat{s}_1, \ldots, \hat{s}_T$.

Example 2.1

Continue Example 1.4. Given the observed workload data sequence $o_{1:T}$, the MAP estimation of the hidden states can be obtained using Eqn (2.15), which are partially listed in the following table. It shows that the process often stays in a state for a long period of time. The average duration is 87.8 s and the maximum duration 405 s. Figure 1.7 shows the data trace and the estimated hidden states. The hidden states are represented by the arrival rates in the plot.

$\hat{s}$	11	6	9	18	7	10	16	9	12	8	10	14
$\hat{d}$	67	21	274	5	112	109	8	154	14	405	87	30

Note that a subsequence of the estimated states $\hat{s}_t = i, \hat{s}_{t+1:t+d} = j, \hat{s}_{t+d+1} = k$, for $i \neq j \neq k$, may not represent a valid state j with duration d, because there may be $P[S_{[t+1:t+d]} = j, o_{1:T} | \lambda] = 0$.

If we choose $\eta_t(i,d)$ of Eqn (2.10), instead of $\gamma_t(i)$, as the MAP criterion, we can obtain a valid MAP estimation of the state that crosses time t given specific observation sequence $o_{1:T}$, that is,

$$(\hat{i}_t, \hat{d}_t, \hat{\tau}_t) = \arg \max_{\substack{i,d,\tau \\ \tau-d+1 \leq t \leq \tau}} \bar{\eta}_\tau(i,d) = \arg \max_{\substack{i,d,\tau \\ \tau-d+1 \leq t \leq \tau}} \eta_\tau(i,d). \tag{2.16}$$

The confidence of the state and duration estimation is

$$\left[\eta_{\hat{\tau}_t}(\hat{i}_t, \hat{d}_t) - \max_{\substack{i,d,\tau \\ (i,d)\neq(\hat{i}_t,\hat{d}_t),\, \tau\neq\hat{\tau}_t \\ \tau-d+1\leq t\leq\tau}} \eta_\tau(i,d)\right] / \eta_{\hat{\tau}_t}(\hat{i}_t, \hat{d}_t).$$

However, Eqn (2.16) cannot be used to estimate the state sequence because there may exist inconsistent estimation results. For example, when $\hat{\tau}_t \geq t+1$, $(\hat{i}_t, \hat{d}_t, \hat{\tau}_t)$ means $i_{t+1} = \hat{i}_t$, while the next estimation $(\hat{i}_{t+1}, \hat{d}_{t+1}, \hat{\tau}_{t+1})$ may yield $\hat{i}_{t+1} \neq \hat{i}_t$ or $\hat{d}_{t+1} \neq \hat{d}_t$.

Besides, one may use sequential estimation method to get the state sequence. That is, use Eqn (2.16) to get $(\hat{i}_T, \hat{d}_T, \hat{\tau}_T)$ at first, and then let $t = \hat{\tau}_T - \hat{d}_T$ and

$$(\hat{i}_t, \hat{d}_t) = \arg\max_{i,d} \overline{\eta}_t(i,d) = \arg\max_{i,d} \eta_t(i,d). \tag{2.17}$$

After that, let $t = t - \hat{d}_t$ and use Eqn (2.17) again, ..., until $t \leq 0$. Finally, a state sequence is obtained. However, it cannot be proved that this state sequence is the best.

2.3 MATRIX EXPRESSION OF THE FORWARD–BACKWARD ALGORITHM

One usually uses matrix-vector calculations to implement the forward–backward algorithms. To benefit from the matrix computation capability, it is better to express the forward–backward algorithms using matrices. To express the forward algorithm using matrices, the forward formula (2.6) can be changed as

$$\alpha_{t+d}(j,d) = \sum_{i\in\mathbf{S}\setminus\{j\}} \sum_{h\in\mathbf{D}} \alpha_t(i,h) \cdot a_{(i,h)(j,d)} \cdot b_{j,d}(o_{t+1:t+d}). \tag{2.18}$$

Let $\mathbf{A}$ be a $MD \times MD$ state transition probability matrix with elements $a_{(i,h)(j,d)}$, for $i,j \in \mathbf{S}$ and $h,d \in \mathbf{D}$, that is,

$$\mathbf{A} = \begin{bmatrix} a_{(1,1)(1,1)} & a_{(1,1)(1,2)} & \cdots & a_{(1,1)(M,D)} \\ a_{(1,2)(1,1)} & a_{(1,2)(1,2)} & \cdots & a_{(1,2)(M,D)} \\ \vdots & \vdots & \ddots & \vdots \\ a_{(M,D)(1,1)} & a_{(M,D)(1,2)} & \cdots & a_{(M,D)(M,D)} \end{bmatrix},$$

and $\mathbf{B}_{t+1}$ be a $MD \times MD$ diagonal matrix with the (j,d)th diagonal element being observation probability $b_{j,d}(o_{t+1:t+d})$, for $j \in \mathbf{S}$ and $d \in \mathbf{D}$, that is,

$$\mathbf{B}_{t+1} = \begin{bmatrix} b_{1,1}(o_{t+1:t+1}) & 0 & \cdots & 0 \\ 0 & b_{1,2}(o_{t+1:t+2}) & \cdots & 0 \\ \vdots & \vdots & \ddots & \vdots \\ 0 & 0 & \cdots & b_{M,D}(o_{t+1:t+D}) \end{bmatrix},$$

where the elements of the matrices are arranged in the following lexicographic order, that is, *order* $= (j-1)D + d$:

subscript (j,d)	(1,1)	...	(1,*D*)	(2,1)	...	(2,*D*)	...	(*M*,1)	...	(*M*,*D*)
order	1	...	D	$D+1$	...	$2D$	...	$(M-1)D+1$	...	MD

For simplicity, we use row vectors to express the forward variables and column vectors for the backward variables. Let $\vec{\alpha}_t(i)$ be a $1 \times D$ vector with the hth element being $\alpha_t(i,h)$, for $h \in \mathbf{D}$, that is,

$$\vec{\alpha}_t(i) = (\alpha_t(i,1), \alpha_t(i,2), \ldots, \alpha_t(i,D)),$$

for $i \in \mathbf{S}$, and

$$\vec{\alpha}_t = (\vec{\alpha}_t(1), \ldots, \vec{\alpha}_t(M))$$

be a $1 \times MD$ vector of the forward variables that the states end at t. Let $\vec{\alpha}^*_{t+1}(j)$ be a $1 \times D$ vector with the dth element being $\alpha_{t+d}(j,d)$, for $d \in \mathbf{D}$, that is,

$$\vec{\alpha}^*_{t+1}(j) = (\alpha_{t+1}(j,1), \alpha_{t+2}(j,2), \ldots, \alpha_{t+D}(j,D)),$$

for $j \in \mathbf{S}$, and

$$\vec{\alpha}^*_{t+1} = (\vec{\alpha}^*_{t+1}(1), \ldots, \vec{\alpha}^*_{t+1}(M)) \tag{2.19}$$

be a $1 \times MD$ vector of the forward variables that the states start at $t+1$. Then, the forward formula (2.18) becomes

$$\vec{\alpha}^*_{t+1} = \vec{\alpha}_t \mathbf{A} \mathbf{B}_{t+1}. \tag{2.20}$$

In other words, the vector of forward variables that the states start at $t+1$ is transited from the vector of forward variables that the states ended at t.

Now we derive the relationship between $\vec{\alpha}_t(i)$ and $\vec{\alpha}_t^*(i)$. Let $\mathbf{E}_d$ be a $MD \times MD$ matrix with the $((j-1)D+d)$th diagonal element being 1 and all other elements being 0, for $j = 1, \ldots, M$ and given d. In other words, $\mathbf{E}_d$ contains M nonzero elements. Obviously, $\sum_{d \in \mathbf{D}} \mathbf{E}_d = \mathbf{E}$, where $\mathbf{E}$ is an $MD \times MD$ identity matrix. For simplicity, we let $\mathbf{E}_{D+1} = 0$. Then $\vec{\alpha}_t = \vec{\alpha}_t^* \mathbf{E}_1 + \cdots + \vec{\alpha}_{t-D+1}^* \mathbf{E}_D$. Therefore, the forward formula (2.20) can be rewritten as

$$\begin{aligned} \vec{\alpha}_t &= (\vec{\alpha}_t^* \mathbf{E}_1 + \cdots + \vec{\alpha}_{t-D+1}^* \mathbf{E}_D) = \sum_{d=1}^{D} \vec{\alpha}_{t-d+1}^* \mathbf{E}_d \\ &= \sum_{d=1}^{D} \vec{\alpha}_{t-d} \mathbf{A} \mathbf{B}_{t-d+1} \mathbf{E}_d \end{aligned} \tag{2.21}$$

or

$$\vec{\alpha}_{t+1}^* = \sum_{d=1}^{D} \vec{\alpha}_{t-d+1}^* \mathbf{E}_d \mathbf{A} \mathbf{B}_{t+1}.$$

Under the general assumptions that the first state can start at or before $t = 1$, the initial values of the forward vectors are

$$\begin{aligned} \vec{\alpha}_\tau &= (\alpha_\tau(1,1), \ldots, \alpha_\tau(1,D), \ldots, \alpha_\tau(M,1), \ldots, \alpha_\tau(M,D)) \\ &= (\pi_{1,1}, \pi_{1,2}, \ldots, \pi_{M,D}) = \vec{\pi}, \end{aligned}$$

for $\tau \leq 0$, and from Eqn (2.20)

$$\vec{\alpha}_\tau^* = \vec{\pi} \mathbf{A} \mathbf{B}_\tau,$$

for $\tau \leq 1$.

Similarly, let $\vec{\beta}_t(j)$ be a $D \times 1$ vector with the dth element being $\beta_t(j,d)$, for $d \in \mathbf{D}$, that is, for $j \in \mathbf{S}$,

$$\vec{\beta}_t(j) = \begin{bmatrix} \beta_t(j,1) \\ \beta_t(j,2) \\ \vdots \\ \beta_t(j,D) \end{bmatrix}, \quad \text{and} \quad \vec{\beta}_t = \begin{bmatrix} \vec{\beta}_t(1) \\ \vec{\beta}_t(2) \\ \vdots \\ \vec{\beta}_t(M) \end{bmatrix}$$

be a $MD \times 1$ vector of backward variables that the given states end at t. Let $\vec{\beta}^*_{t+1}(j)$ be a $D \times 1$ vector with the dth element being $\beta_{t+d}(j,d)$, for $d \in \mathbf{D}$, that is, for $j \in \mathbf{S}$

$$\vec{\beta}^*_{t+1}(j) = \begin{bmatrix} \beta_{t+1}(j,1) \\ \beta_{t+2}(j,2) \\ \vdots \\ \beta_{t+D}(j,D) \end{bmatrix}, \quad \text{and} \quad \vec{\beta}^*_{t+1} = \begin{bmatrix} \vec{\beta}^*_{t+1}(1) \\ \vec{\beta}^*_{t+1}(2) \\ \vdots \\ \vec{\beta}^*_{t+1}(M) \end{bmatrix} \tag{2.22}$$

be a $MD \times 1$ vector of backward variables that the given states start at $t+1$. Then, the backward formula (2.7) becomes

$$\vec{\beta}_t = \mathbf{AB}_{t+1}\vec{\beta}^*_{t+1}. \tag{2.23}$$

In other words, the vector of backward variables that the given states start at $t+1$ is transited from the vector of backward variables that the given states end at t.

Because $\vec{\beta}^*_t = \mathbf{E}_1\vec{\beta}_t + \cdots + \mathbf{E}_D\vec{\beta}_{t+D-1} = \sum_{d=1}^{D} \mathbf{E}_d\vec{\beta}_{t+d-1}$, the backward formula (2.23) can be rewritten as

$$\vec{\beta}_t = \mathbf{AB}_{t+1}\sum_{d=1}^{D} \mathbf{E}_d\vec{\beta}_{t+d} \tag{2.24}$$

or

$$\vec{\beta}^*_t = \sum_{d=1}^{D} \mathbf{E}_d\mathbf{AB}_{t+d}\vec{\beta}^*_{t+d}. \tag{2.25}$$

Under the general assumptions that the last state is not necessarily ending at $t = T$, the initial values for the backward variables are $\beta_\tau(j,d) = 1$ for $\tau \geq T$. Let $\vec{e}$ be an all-ones $MD \times 1$ vector. Then $\vec{\beta}_\tau = \vec{e}$ and $\vec{\beta}^*_\tau = \sum_{d=1}^{D} \mathbf{E}_d\vec{\beta}_{\tau+d-1} = \vec{e}$, for $\tau \geq T$.

Therefore, the matrix-form forward–backward algorithm for the general HSMM is as follows:

Algorithm 2.2 Matrix-Form Forward–Backward Algorithm for the General HSMM

The Matrix-Form Forward Algorithm

1. *If the simplified boundary condition is assumed* {

  $\vec{\alpha}_0 = \vec{\pi}$ *and* $\vec{\alpha}_\tau = 0$ *for* $\tau < 0$; *or*

  $\vec{\alpha}_1^* = \vec{\pi}\mathbf{A}\mathbf{B}_1 = \vec{\Pi}\mathbf{B}_1$ *and* $\vec{\alpha}_\tau^* = 0$ *for* $\tau \le 0$;

 } *else* {

  $\vec{\alpha}_\tau = \vec{\pi}$ *for* $\tau \le 0$; *or* $\vec{\alpha}_\tau^* = \vec{\pi}\mathbf{A}\mathbf{B}_\tau$ *for* $\tau \le 1$;

 }

2. *For* $t = 1, \ldots, T$ {

  $\vec{\alpha}_t = \sum_{d=1}^{D} \vec{\alpha}_{t-d}\mathbf{A}\mathbf{B}_{t-d+1}\mathbf{E}_d$;

 } *or*

 For $t = 2, \ldots, T$ {

  $\vec{\alpha}_t^* = \sum_{d=1}^{D} \vec{\alpha}_{t-d}^*\mathbf{E}_d\mathbf{A}\mathbf{B}_t$.

 }

The Matrix-Form Backward Algorithm

1. *If the simplified boundary condition is assumed* {

  $\vec{\beta}_T = \vec{e}$ *and* $\vec{\beta}_\tau = 0$ *for* $\tau > T$; *or*

  $\vec{\beta}_T^* = \mathbf{E}_1\vec{e}$, $\vec{\beta}_\tau^* = 0$ *for* $\tau > T$;

 } *else* {

  $\vec{\beta}_\tau = \vec{e}$ *for* $\tau \ge T$; *or*

  $\vec{\beta}_\tau^* = \vec{e}$ *for* $\tau \ge T$;

 }

2. *For* $t = T - 1, \ldots, 1$ {

  $\vec{\beta}_t = \mathbf{A}\mathbf{B}_{t+1}\sum_{d=1}^{D} \mathbf{E}_d\vec{\beta}_{t+d}$; *or*

  $\vec{\beta}_t^* = \sum_{d=1}^{D} \mathbf{E}_d\mathbf{A}\mathbf{B}_{t+d}\vec{\beta}_{t+d}^*$.

 }

Note that in computing the forward vectors, the observations considered may cross the boundary $t = 1$. That is, when $1 - \tau < 1$ and $d - \tau \ge 1$, observations $o_{1-\tau:d-\tau}$ contains unobserved values $o_{1-\tau:0}$. In this case,

$$b_{j,d}(o_{[1-\tau:d-\tau]}) = \sum_{v_{k_1:k_\tau}} b_{j,d}(v_{k_1:k_\tau} o_{1:d-\tau]}) = b_{j,d}(o_{1:d-\tau]}),$$

where $o_{1:d-\tau]}$ denotes the last substring, with length $d-\tau$, of an observable string $v_{k_1:k_d}$ of length d for given state (j,d). In computing the backward vectors, when $T-\tau+1 \le T$ and $T-\tau+d > T$, the observations $o_{T-\tau+1:T-\tau+d}$ cross the boundary $t=T$ with unobserved observations $o_{T+1:T-\tau+d}$. In this case,

$$b_{j,d}(o_{[T-\tau+1:T-\tau+d]}) = \sum_{v_{k_{\tau+1}:k_d}} b_{j,d}(o_{[T-\tau+1:T}v_{k_{\tau+1}:k_d}) = b_{j,d}(o_{[T-\tau+1:T}),$$

where $o_{[T-\tau+1:T}$ denotes the first substring, with length τ, of an observable string $v_{k_1:k_d}$ of length d for given state (j,d).

2.4 FORWARD-ONLY ALGORITHM FOR HSMM

Because the backward recursion is required, the forward–backward algorithms are not suited for very long observation sequences or online calculation when the length of observation sequence is continuously increasing. Therefore, a forward-only algorithm of HSMM is essential. A forward-only algorithm can be extremely complex using the sum-product form. However, it can be relatively simple if it is expressed using matrix calculation.

2.4.1 A General Forward-Only Algorithm

For inference and estimation of HSMM, the variables that are required to compute include the forward variables $\alpha_t(j,d)$, the backward variables $\beta_t(j,d)$, the smoothed probabilities $\overline{\eta}_t(j,d)$, $\overline{\xi}_t(i,h;j,d)$, $\overline{\gamma}_t(j)$, and the expectations $\sum_t \overline{\eta}_t(j,d)$, $\sum_t \overline{\xi}_t(i,h;j,d)$, $\sum_t \overline{\gamma}_t(j)$. The forward-only algorithm is thus aiming at computing all the variables $\alpha_t(j,d)$, $\eta_t(j,d)$, $\xi_t(i,h;j,d)$, $\gamma_t(j)$, $\sum_t \eta_t(j,d)$, $\sum_t \xi_t(i,h;j,d)$, and $\sum_t \gamma_t(j)$ without computing the backward variables $\beta_t(j,d)$. From the matrix form of the backward formula (2.24), we can see that the backward variables $\vec{\beta}_t$ can be transformed into $\mathbf{AB}_{t+1} \sum_{d=1}^{D} \mathbf{E}_d \vec{\beta}_{t+d}$ that the backward variables have higher indices $t+d>t$. When the indices become greater than T, the initial conditions that $\vec{\beta}_t = 1$, for $t \ge T$, can be applied. Therefore, this makes the forward-only algorithm available.

Now we derive the forward-only algorithm. Suppose a vector of smoothed probabilities generally has the following form:

$$\vec{\varphi}_t = \mathbf{Y}_t \vec{\beta}_t,$$

where $\vec{\beta}_t$ is the vector of the backward variables. For example, let $\vec{\eta}_t = (\eta_t(1,1), \ldots, \eta_t(1,D), \ldots, \eta_t(M,1), \ldots, \eta_t(M,D))^T$ and $Diag(\vec{\alpha}_t)$ be a $MD \times MD$ diagonal matrix with the $((j-1)D+d)$th diagonal element being the $((j-1)D+d)$th element of $\vec{\alpha}_t$. Then $\vec{\eta}_t = Diag(\vec{\alpha}_t)\vec{\beta}_t$. In this case, $\vec{\varphi}_t = \vec{\eta}_t$ and $\mathbf{Y}_t = Diag(\vec{\alpha}_t)$. Hence, a vector of expectations of parameters has the general form

$$\sum_{t \le T} \vec{\varphi}_t = \sum_{t \le T} \mathbf{Y}_t \vec{\beta}_t.$$

For example, the vector of expectations that state j has duration d, for all j and d, are $\sum_{t \le T} \vec{\eta}_t$. Besides, for given $t_0 \ll T$, the smoothed probabilities $\vec{\eta}_{t_0}(j,d)$ or $\eta_{t_0}(j,d) = \alpha_{t_0}(j,d)\beta_{t_0}(j,d)$ can also be expressed in this form to avoid computing the backward variables from T back to t_0. That is, let $\vec{\varphi}_{t_0} = \sum_{t_0 \le t \le T} \mathbf{Y}_t \vec{\beta}_t$ and $\mathbf{Y}_t = 0$ for $t > t_0$.

To iteratively compute the sum of the vectors, $\sum_{t \le T} \vec{\varphi}_t$, we assume

$$\sum_{\tau \le t} \vec{\varphi}_\tau = \sum_{d=1}^{D} \mathbf{X}_t(d) \vec{\beta}_{t+d-1},$$

where $\mathbf{X}_t(d)$ are the forward matrices to be computed. Let $\mathbf{B}'_{t+d} = \mathbf{B}_{t+1}\mathbf{E}_d$, which contains observations $o_{t+1:t+d}$. Since $\vec{\beta}_t = \mathbf{A} \sum_{d=1}^{D} \mathbf{B}'_{t+d} \vec{\beta}_{t+d}$,

$$\begin{aligned}
\sum_{\tau \le t+1} \vec{\varphi}_\tau &= \sum_{d=1}^{D} \mathbf{X}_t(d) \vec{\beta}_{t+d-1} + \mathbf{Y}_{t+1} \vec{\beta}_{t+1} \\
&= \sum_{d=2}^{D} \mathbf{X}_t(d) \vec{\beta}_{t+d-1} + \mathbf{A} \sum_{d=1}^{D} \mathbf{B}'_{t+d} \vec{\beta}_{t+d} + \mathbf{Y}_{t+1} \vec{\beta}_{t+1} \\
&= \sum_{d=1}^{D-1} \mathbf{X}_t(d+1) \vec{\beta}_{t+d} + \mathbf{A} \sum_{d=1}^{D} \mathbf{B}'_{t+d} \vec{\beta}_{t+d} + \mathbf{Y}_{t+1} \vec{\beta}_{t+1}.
\end{aligned}$$

Let

$$\begin{aligned}
&\mathbf{X}_{t+1}(1) = \mathbf{X}_t(2) + \mathbf{A}\mathbf{B}'_{t+1} + \mathbf{Y}_{t+1} \\
&\mathbf{X}_{t+1}(d) = \mathbf{X}_t(d+1) + \mathbf{A}\mathbf{B}'_{t+d}, \quad 1 < d < D \qquad (2.26) \\
&\mathbf{X}_{t+1}(D) = \mathbf{A}\mathbf{B}'_{t+D}.
\end{aligned}$$

be the forward-only iterations. Then $\sum_{t\leq T}\vec{\varphi}_t$ can be determined iteratively. Finally, when $\mathbf{X}_T(d)$ for $d=1,\ldots,D$ are determined, we can apply the initial conditions for the backward variables $\vec{\beta}_\tau=\vec{e}$ for $\tau\geq T$, and get

$$\sum_{t\leq T}\vec{\varphi}_t=\sum_{d=1}^{D}\mathbf{X}_T(d)\,\vec{e}\,. \tag{2.27}$$

Note that $\mathbf{B}'_{t+d}$ contains observations $o_{t+1:t+d}$. When $t+d>T$, the future observations $o_{T+1:t+d}$ are unknown, that is, we must let its elements $b_{j,d}(o_{t+1:t+d})=b_{j,d}(o_{[t+1:T})$, where $o_{[t+1:T}$ denotes the first part of the observable strings $o_{[t+1:T+d]}$ of length d. Once T is increased, the newly observed o_{T+1} were not contained in $\mathbf{B}'_{t+d}$. Therefore, Eqn (2.27) is applicable for given length T but not for increasing T. To let Eqn (2.27) be valid for increasing T, the matrices $\mathbf{X}_{t+1}(d)$, for $t+d>T$, cannot be used for future iterations, and must be recomputed at the next length $T+1$ to include the newly obtained observation o_{T+1}. In summary, the forward-only algorithm is as follows:

Algorithm 2.3 Forward-Only Algorithm

1. *Let* $\mathbf{X}_1(1)=\mathbf{Y}_1$ *and* $\mathbf{X}_1(d)=0$ *for* $d>1$; *Let* $\mathrm{T}=1$;
2. *From* $\mathbf{Y}_{T-1}$ *and* $o_{T-D+1:T}$ *compute* $\mathbf{Y}_T$;
3. *For* $t=T-D,\ldots,T-1$ {
   For $d=1,\ldots,D$ {
   If $t+d\geq T${
   Use observations $o_{[t+1:T}$ *to construct* $\mathbf{B}'_{t+d}$;
   Use Eqn (2.26) *to compute* $\mathbf{X}_{t+1}(d)$;
   } } }
4. *Compute the expectation vector* $\vec{g}_T=\sum_{t\leq T}\vec{\varphi}_t=\sum_{d=1}^{D}\mathbf{X}_T(d)\vec{e}$;
5. $\mathrm{T}=\mathrm{T}+1$; *go back to step 2.*

For simplicity, we denote the result of the forward-only iteration by $\vec{g}_T=\sum_{t\leq T}\vec{\varphi}_t=\vec{G}\left(\sum_{t\leq T}\mathbf{Y}_t\vec{\beta}_t\right)$.

2.4.2 Computing Smoothed Probabilities and Expectations

Suppose the set of observable substrings of length d is $\mathbf{V}_d=\{v_{d,1},v_{d,2},\ldots,v_{d,L_d}\}$, where L_d is total number of the observable substrings of length d, for $d\in\mathbf{D}$. Let $L=\max_d\{L_d\}$ and $b_{j,d}(v_{d,l})=0$ when $l>L_d$ or $v_{d,l}\notin\mathbf{V}_d$. Now we apply the forward-only algorithm to online compute the expectations $\sum_{t=1}^{T} \bar{\eta}_t(j, d) I(o_{t-d+1:t} = v_{d,l})$, where the indicator function $I(x) = 1$ if x is true and 0 otherwise, and $I(o_{t-d+1:t} = v_{d,l}) = 0$ if $t - d + 1 < 1$ or $v_{d,l} \notin \mathbf{V}_d$. This vector of expectations are usually used for estimation of the observation distributions.

Construct a $MD \times MD$ diagonal matrix of indicators: $\mathbf{I}_t(l) = \sum_{d=1}^{D} I(o_{t-d+1:t} = v_{d,l}) \mathbf{E}_d$, for $l = 1, \ldots, L$. Then, let $\mathbf{I}_t = \begin{bmatrix} \mathbf{I}_t(1) \\ \vdots \\ \mathbf{I}_t(L) \end{bmatrix}$ be a $MDL \times MD$ indicator matrix. The expectation vector for the observation/emission probabilities is thus $\sum_{t=1}^{T} \mathbf{I}_t \vec{\eta}_t = \sum_{t=1}^{T} \mathbf{I}_t Diag(\vec{\alpha}_t) \vec{\beta}_t$.

Using the forward-only algorithm of Algorithm 2.3, the expectation vector

$$\vec{g}_T^{\mathbf{B}} = \sum_{t=1}^{T} \mathbf{I}_t \vec{\eta}_t = \vec{G} \left(\sum_{t=1}^{T} [\mathbf{I}_t Diag(\vec{\alpha}_t)] \vec{\beta}_t \right)$$

can be determined. Let $k = MD(l-1) + (j-1)D + d$. Then $b_{j,d}(v_{d,l})$ is corresponding to $\vec{g}_T^{\mathbf{B}}(k)$, for $j = 1, \ldots, M$, $d = 1, \ldots, D$, and $l = 1, \ldots, L$.

Now we consider how to online compute the expectations $\sum_{t=h}^{T-d} \xi_t(i, h; j, d)$, which are usually used for estimation of the state transition probability matrix $\mathbf{A}$. Set $\alpha_t(i, h) = 0$ for $t < h$ and $\beta_t(j, d) = 0$ for $t > T$. Since $\xi_t(i, h; j, d) = \alpha_t(i, h) a_{(i,h)(j,d)} b_{j,d}(o_{t+1:t+d}) \beta_{t+d}(j, d)$, we have $\xi_t(i, h; j, d) = 0$ for $t < h$ and $t > T - d$. The expectations are $\sum_{t=1}^{T} \xi_t(i, h; j, d)$.

Define $1 \times MD$ vectors $\vec{e}_{i,h} = (0, \ldots, 1, \ldots, 0)$ with unity as the (i,h)'th element and zeros elsewhere. Let $\mathbf{Z}_t(i, h) = Diag(\vec{e}_{i,h} Diag(\vec{\alpha}_t) \mathbf{A})$ be a $MD \times MD$ diagonal matrix with the diagonal elements being the elements of the (i,h)'th row of $Diag(\vec{\alpha}_t)\mathbf{A}$, that is, $\alpha_t(i, h) a_{(i,h)(j,d)}$ for $j = 1, \ldots, M$, $d = 1, \ldots, D$, and

$$\mathbf{Z}_t = \begin{bmatrix} \mathbf{Z}_t(1,1) \\ \mathbf{Z}_t(1,2) \\ \vdots \\ \mathbf{Z}_t(M,D) \end{bmatrix}$$

be a $(MD)^2 \times MD$ matrix. Therefore, $\mathbf{Z}_t \mathbf{B}_{t+1} \vec{\beta}^*_{t+1}$ is a $(MD)^2 \times 1$ vector. Its $[(i-1)MD^2 + (h-1)MD + (j-1)D + d]$th element is $\xi_t(i,h;j,d)$, for $i,j \in \mathbf{S}$ and $h,d \in \mathbf{D}$. Since $\vec{\beta}^*_{t+1} = \sum_{d=1}^{D} \mathbf{E}_d \vec{\beta}_{t+d}$, the expectations $\sum_{t=1}^{T} \xi_t(i,h;j,d)$, for $i,j \in \mathbf{S}$ and $h,d \in \mathbf{D}$, can be expressed using the following matrix:

$$\sum_{t=1}^{T}\sum_{d=1}^{D} \mathbf{Z}_t \mathbf{B}_{t+1} \mathbf{E}_d \vec{\beta}_{t+d} = \sum_{d=1}^{D}\sum_{t=d+1}^{T+d} \mathbf{Z}_{t-d} \mathbf{B}_{t-d+1} \mathbf{E}_d \vec{\beta}_t$$

$$= \sum_{d=1}^{D}\sum_{t=d+1}^{T} \mathbf{Z}_{t-d} \mathbf{B}_{t-d+1} \mathbf{E}_d \vec{\beta}_t$$

$$= \sum_{t=2}^{T}\sum_{d=1}^{\min\{D,t-1\}} \mathbf{Z}_{t-d} \mathbf{B}_{t-d+1} \mathbf{E}_d \vec{\beta}_t.$$

Using the forward-only algorithm of Algorithm 2.3, the expectation vector of the state transition probabilities $\vec{g}_T^{\mathbf{A}} = \vec{G}\left(\sum_{t=2}^{T}\left(\sum_{d=1}^{\min\{D,t-1\}} \mathbf{Z}_{t-d} \mathbf{B}_{t-d+1} \mathbf{E}_d\right)\vec{\beta}_t\right)$ can be determined. Let $k = (i-1)MD^2 + (h-1)MD + (j-1)D + d$. Then $a_{(i,h)(j,d)}$ is corresponding to $\vec{g}_T^{\mathbf{A}}(k)$, for $i,j \in \mathbf{S}$ and $h,d \in \mathbf{D}$.

Finally we can apply the forward-only algorithm of Algorithm 2.3 to compute the smoothed probabilities $\overline{\eta}_d(j,d)$ of the starting states, which are usually used for estimation of the initial state distribution. Let an $MD \times 1$ vector $(\eta_1(1,1),\ldots,\eta_D(1,D),\ldots,\eta_1(M,1),\ldots,\eta_D(M,D))^T = Diag(\vec{\alpha}^*_1)\vec{\beta}^*_1$. Since $\vec{\alpha}^*_1 = \vec{\Pi}\mathbf{B}_1$ and $\vec{\beta}^*_1 = \sum_{d=1}^{D} \mathbf{E}_d \vec{\beta}_d$, we have $Diag(\vec{\alpha}^*_1)\vec{\beta}^*_1 = \sum_{d=1}^{D} Diag(\vec{\Pi}\mathbf{B}_1)\mathbf{E}_d \vec{\beta}_d$. Let $\mathbf{Y}_d = Diag(\vec{\Pi}\mathbf{B}_1)\mathbf{E}_d$, for $d = 1,\ldots,D$, and $\mathbf{Y}_t = 0$, for $t > D$. Then the smoothed probabilities of the starting states $\vec{g}_T^{\Pi} = \vec{G}\left(\sum_{t=1}^{T} \mathbf{Y}_t \vec{\beta}_t\right)$ can be determined.

To compute the smoothed probabilities $\overline{\gamma}_t(j)$, define a vector

$$\vec{\gamma}_t = (\gamma_t(1),\ldots,\gamma_t(M))^T,$$

where

$$\gamma_t(j) = \sum_{\substack{\tau, d \\ \tau \geq t \geq \tau - d + 1}} P[S_{[\tau-d+1:\tau]} = j, o_{1:T} | \lambda]$$

$$= \sum_{\tau=t}^{t+D-1} \sum_{d=\tau-t+1}^{D} \alpha_\tau(j,d)\beta_\tau(j,d).$$

Let $1 \times D$ vectors $\vec{c}_d = (0, ..., 0, 1, ..., 1)$ contain d zeros and $D - d$ ones, and

$$\mathbf{C}_d = \begin{bmatrix} \vec{c}_d & 0 & \cdots & 0 \\ 0 & \vec{c}_d & \ddots & \vdots \\ \vdots & \ddots & \ddots & 0 \\ 0 & \cdots & 0 & \vec{c}_d \end{bmatrix},$$

as an $M \times MD$ matrix. Then

$$\vec{\gamma}_t = \sum_{\tau=t}^{t+D-1} [\mathbf{C}_{\tau-t} Diag(\vec{\alpha}_\tau)]\vec{\beta}_\tau. \tag{2.28}$$

Therefore, $\vec{\gamma}_t$ can be computed by letting $\mathbf{Y}_\tau = \mathbf{C}_{\tau-t} Diag(\vec{\alpha}_\tau)$ for $\tau = t, \ldots, t + D - 1$, and $\mathbf{Y}_\tau = 0$ for $\tau \geq t + D$. Using the forward-only algorithm of Algorithm 2.3, $\vec{\gamma}_t = \vec{G}\left(\sum_{\tau=t}^{T} \mathbf{Y}_\tau \vec{\beta}_\tau\right)$. After $\vec{\gamma}_t$ is determined for any t, the likelihood of the observations can be determined by

$$P[o_{1:T} | \lambda] = \vec{e}^T \vec{\gamma}_t.$$

2.5 VITERBI ALGORITHM FOR HSMM

Viterbi algorithm for HMM is a well-known algorithm for finding the most likely sequence of states. There exist similar algorithms for the HSMM (Ljolje and Levinson, 1991; Ramesh and Wilpon, 1992; Chen et al., 1995; Burshtein, 1996). We will call it *Viterbi HSMM algorithm* in this book to distinguish it from the conventional Viterbi algorithm.

Let $S_{1:T}$ be a state sequence corresponding to the observation sequence $o_{1:T}$. It is to find an instance $s_{1:T}$ such that $P[s_{1:T} | o_{1:T}, \lambda]$ is maximized. Define the forward variables for the Viterbi HSMM algorithm by

$$\begin{aligned}\delta_t(j,d) &\equiv \max_{S_{1:t-d}} P[S_{1:t-d}, S_{[t-d+1:t]}=j, o_{1:t}|\lambda] \\ &= \max_{i\neq j}\max_{h}\max_{S_{1:t-d-h}} P[S_{1:t-d-h}, S_{[t-d-h+1,t-d]}=i, S_{[t-d+1:t]}=j, o_{1:t}|\lambda] \\ &= \max_{i\neq j}\max_{h}\{\delta_{t-d}(i,h)a_{(i,h)(j,d)}b_{j,d}(o_{t-d+1:t})\},\end{aligned} \tag{2.29}$$

for $1\le t\le T,\ j\in\mathbf{S}, d\in\mathbf{D}$. $\delta_t(j,d)$ represents the most likely partial state sequence that ends at t in state j of duration d. Record the previous state that $\delta_t(j,d)$ selects by $\Psi(t,j,d)\equiv(t-d,i^*,h^*)$, where i^* is the previous state survived, h^* its duration, and $(t-d)$ its ending time. $\Psi(t,j,d)$ is determined by letting

$$(i^*,h^*) = \arg\max_{i\in\mathbf{S}\backslash\{j\}}\max_{h\in\mathbf{D}}\{\delta_{t-d}(i,h)a_{(i,h)(j,d)}b_{j,d}(o_{t-d+1:t})\}.$$

Now the state sequence can be determined by finding the last state that maximizes the likelihood. For the general assumption of the boundary conditions, the last state is

$$(t_1,j_1^*,d_1^*) = \arg\max_{T+D-1\ge t\ge T}\max_{j\in\mathbf{S}}\max_{D\ge d\ge t-T+1}\delta_t(j,d),$$

or, for the simplified assumption of the boundary conditions, $t_1=T$ and

$$(j_1^*,d_1^*) = \arg\max_{j\in\mathbf{S}}\max_{d\in\mathbf{D}}\delta_T(j,d).$$

Trace back the state sequence by letting

$$(t_2,j_2^*,d_2^*) = \Psi(t_1,j_1^*,d_1^*),$$

$$\ldots$$

$$(t_n,j_n^*,d_n^*) = \Psi(t_{n-1},j_{n-1}^*,d_{n-1}^*),$$

until the first state S_1 is determined, where $S_1=j_n^*$ and $(j_n^*,d_n^*),\ldots,(j_1^*,d_1^*)$ is the estimated state sequence.

Algorithm 2.4 Viterbi HSMM Algorithm

The Forward Algorithm

1. *For* $j = 1, \ldots, M$ *and* $d = 1, \ldots, D$, *let* $\delta_0(j,d) = \pi_{j,d}$;
2. *If it is assumed that the first state must start at* $t = 1$
   let $\delta_\tau(j,d) = 0$ *for* $\tau < 0$;
   otherwise
   let $\delta_\tau(j,d) = \pi_{j,d}$ *for* $\tau < 0$;
3. *For* $t = 1, \ldots, T + D - 1$ {
   for $j = 1, \ldots, M$ *and* $d = 1, \ldots, D$ {

$$\delta_t(j,d) = \max_{i \in \mathbf{S}\backslash\{j\}, h \in \mathbf{D}} \{\delta_{t-d}(i,h) a_{(i,h)(j,d)} b_{j,d}(o_{t-d+1:t})\};$$

$$(i^*, h^*) = \arg \max_{i \in \mathbf{S}\backslash\{j\}, h \in \mathbf{D}} \{\delta_{t-d}(i,h) a_{(i,h)(j,d)} b_{j,d}(o_{t-d+1:t})\};$$

$$\Psi(t,j,d) = (t-d, i^*, h^*);$$

   }
   }

The Trace Back Algorithm

1. *If it is assumed that the last state must end at* $t = T$
   let $t_1 = T$ *and* $(j_1^*, d_1^*) = \operatorname{argmax}_{j \in \mathbf{S}} \max_{d \in \mathbf{D}} \delta_T(j,d)$;
   otherwise
   let $(t_1, j_1^*, d_1^*) = \arg\max_{T+D-1 \geq t \geq T} \max_{j \in \mathbf{S}} \max_{D \geq d \geq t-T+1} \delta_t(j,d)$;
2. *For* $n = 2, 3, \ldots$ {
   $(t_n, j_n^*, d_n^*) = \Psi(t_{n-1}, j_{n-1}^*, d_{n-1}^*)$;
   if $t_n - d_n^* + 1 \leq 1$, *end the procedure;*
   }
3. $(j_n^*, d_n^*), \ldots, (j_1^*, d_1^*)$ *is the estimated state sequence.*

If the state duration density function is log-convex parametric, which is fulfilled by the commonly used parametric functions, Bonafonte et al. (1993) empirically showed that the computational complexity can be reduced to about 3.2 times of the conventional HMM. If the model is a left-right HSMM or the particular state sequence, $i_1, \ldots, i_n$, is given, then only the optimal segmentation of state durations needs to be determined. This is accomplished by simply rewriting Eqn (2.29) as (Levinson et al., 1988, 1989)

$$\delta_t(i_m, d) = \max_{h \in D} \{\delta_{t-d}(i_{m-1}, h) a_{(i_{m-1},h)(i_m,d)} b_{i_m,d}(o_{t-d+1:t})\},$$

for $1 \leq m \leq n$, $1 \leq t \leq T$, $d \in \mathbf{D}$.

2.5.1 Bidirectional Viterbi HSMM Algorithm

Similar to the forward–backward algorithm, Viterbi HSMM algorithm can be extended to be bidirectional (Guedon, 2007). The backward variables for the Viterbi HSMM algorithm can be defined as

$$\begin{aligned}
\varepsilon_t(j,d) &\equiv \max_{S_{t+1:T}} P[S_{t+1:T}, o_{t+1:T} | S_{[t-d+1:t]} = j, \lambda] \\
&= \max_{(i,h), S_{t+h+1:T}} P[S_{[t+1:t+h]} = i, S_{t+h+1:T}, o_{t+1:T} | S_{[t-d+1:t]} = j, \lambda] \\
&= \max_{(i,h)} \{ a_{(j,d)(i,h)} b_{i,h}(o_{t+1:t+h}) \varepsilon_{t+h}(i,h) \}.
\end{aligned} \tag{2.30}$$

The backward variable $\varepsilon_t(j,d)$ is the probability that the process will take the best future state sequence $S_{t+1:T}$ starting from the given state $S_{[t-d+1:t]} = j$. Therefore, the probability that the best state path passes through the given state $S_{[t-d+1:t]} = j$ can be determined by the bidirectional Viterbi HSMM algorithm, that is,

$$\max_{S_{1:t-d}, S_{t+1:T}} P[S_{1:t-d}, S_{[t-d+1:t]} = j, S_{t+1:T}, o_{1:T} | \lambda] = \delta_t(j,d) \varepsilon_t(j,d).$$

2.6 CONSTRAINED-PATH ALGORITHM FOR HSMM

By combining the forward–backward algorithm and the bidirectional Viterbi HSMM algorithm, the following four joint probabilities can be obtained:

$$\begin{aligned}
\sum_{S_{1:t-d}} \sum_{S_{t+1:T}} & P[S_{1:t-d}, S_{[t-d+1:t]} = j, S_{t+1:T}, o_{1:T} | \lambda] \\
&= P[S_{[t-d+1:t]} = j, o_{1:T} | \lambda] \\
&= \alpha_t(j,d) \beta_t(j,d),
\end{aligned}$$

$$\begin{aligned}
\max_{S_{1:t-d}} \max_{S_{t+1:T}} & P[S_{1:t-d}, S_{[t-d+1:t]} = j, S_{t+1:T}, o_{1:T} | \lambda] \\
&= \delta_t(j,d) \varepsilon_t(j,d),
\end{aligned}$$

$$\begin{aligned}
\sum_{S_{1:t-d}} \max_{S_{t+1:T}} & P[S_{1:t-d}, S_{[t-d+1:t]} = j, S_{t+1:T}, o_{1:T} | \lambda] \\
&= \max_{S_{t+1:T}} P[S_{[t-d+1:t]} = j, S_{t+1:T}, o_{1:T} | \lambda] \\
&= \alpha_t(j,d) \varepsilon_t(j,d),
\end{aligned}$$

and

$$\max_{S_{1:t-d}} \sum_{S_{t+1:T}} P[S_{1:t-d}, S_{[t-d+1:t]} = j, S_{t+1:T}, o_{1:T} | \lambda]$$

$$= \max_{S_{1:t-d}} P[S_{1:t-d}, S_{[t-d+1:t]} = j, o_{1:T} | \lambda]$$

$$= \delta_t(j, d)\beta_t(j, d).$$

These joint probabilities can be interpreted as the state paths that pass through the given state $S_{[t-d+1:t]} = j$ for the given observation sequence $o_{1:T}$. For example, the first probablility counts for all the state paths that pass through the given state j during the given period of $t - d + 1$ to t, as shown in Figure 2.1(a). The second is the maximum probability that a path passes through the state j during the period of $t - d + 1$ to t, as shown in Figure 2.1(b). The third is the probability that the best subpath $S^*_{t-d+1:T}$ starts from the state $S_{[t-d+1:t]} = j$, as shown in Figure 2.1(c). The fourth is the probability that the best subpath $S^*_{1:t}$ ends in the state $S_{[t-d+1:t]} = j$, given the entire observations, as shown in Figure 2.1(d).

Further, as shown in Figure 2.2(a), the set $\{\alpha_t(j, d)\beta_t(j, d), j \in \mathbf{S}, d \in \mathbf{D}\}$ counts for all the paths of the trellis, and so is similar to a cut set at t that all the paths pass through.

In contrast, $\{\delta_t(j, d)\varepsilon_t(j, d), j \in \mathbf{S}, d \in \mathbf{D}\}$ counts for all the best paths, which form two opposite trees meeting at t, as shown in Figure 2.2(b), where the virtual roots are S_0 and S_{T+1}, respectively. Obviously, $\arg\max_{j,d} \delta_t(j, d)\varepsilon_t(j, d)$ determines the best state sequence. From these two trees, one can find out an alternative path that is not the best path but its probability $\delta_t(j, d)\varepsilon_t(j, d)$ is close to the maximum, that is, $\max_{j,d} \delta_t(j, d)\varepsilon_t(j, d)$. This is useful for some applications when an alternative choice is required.

The sets $\{\alpha_t(j, d)\varepsilon_t(j, d), j \in \mathbf{S}, d \in \mathbf{D}\}$ and $\{\delta_t(j, d)\beta_t(j, d), j \in \mathbf{S}, d \in \mathbf{D}\}$ are the combinations of Figure 2.2(a) and 2.2(b), as shown in Figure 2.2(c) and 2.2(d). The former uses the history to estimate the initial states of the best subpaths and the latter uses the future observations to smooth the estimation of subpaths.

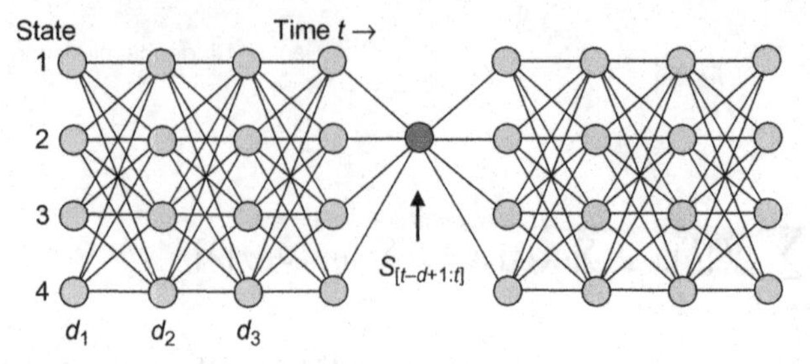

(a) All the paths passing through $S_{[t-d+1:t]} = j$.

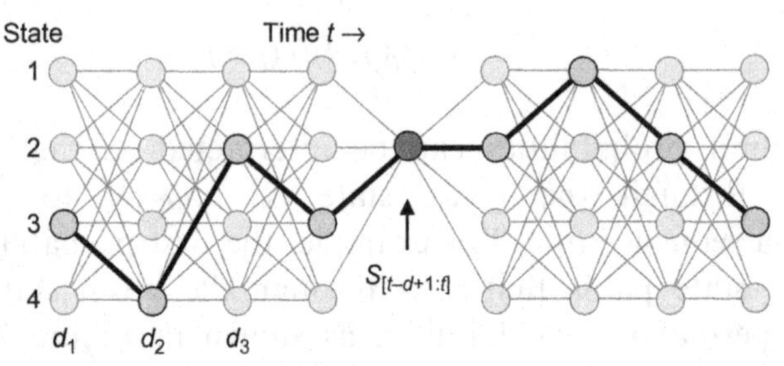

(b) The best path passing through $S_{[t-d+1:t]} = j$.

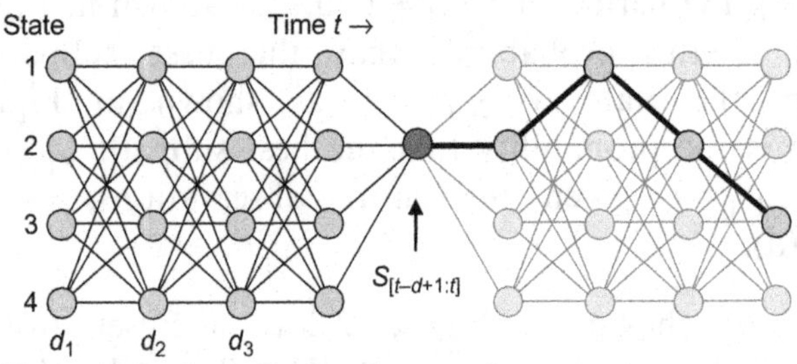

(c) The best subpath starting from $S_{[t-d+1:t]} = j$.

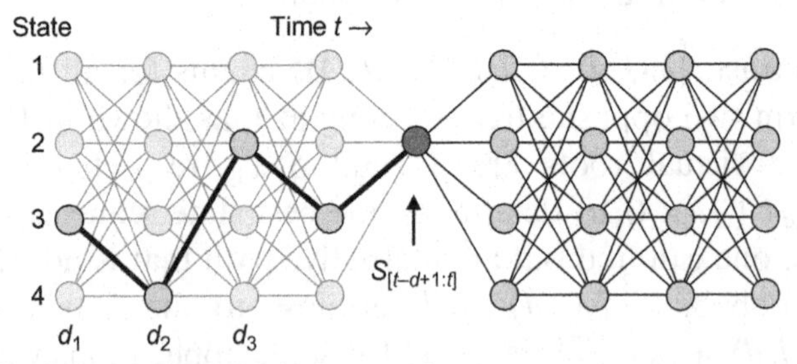

(d) The best subpath ending in $S_{[t-d+1:t]} = j$.

Figure 2.1 Paths passing through a given state.
In the trellis, each circle represents a state and its dwell time, and each line represents a state transition. The time is from left to right. The middle circle represents $s_{[t-d+1:t]} = j$*, which is the given state* j *starting at given time* $t-d+1$ *and ending at given time* t *with given duration* d *for the given observation sequence* $o_{1:T}$*. (a) The previous states from time 1 to the given time* $t-d+1$ *is a random length of state sequence. Each state in a state sequence is a random variable taken from* {1, 2, 3, 4} *with a random duration. The partial trellis from time* 1 *to the given time* $t-d+1$ *represents all such possible paths of states. Similarly, the partial trellis from time* t *to* T *represents all possible future state paths. (b) Among all the previous state paths and the future paths, the black line is the best one that has the maximum likelihood function. Note that if the state path is not constrained passing through the given state* $s_{[t-d+1:t]} = j$*, the best one may not be the black path. (c) If the previous state path is not of concern, which can be any state path, then the best future state path among all others is the black one. (d) Conversely, if the future state path is not of concern, which can be any state path, the black line represents the best previous state path.*

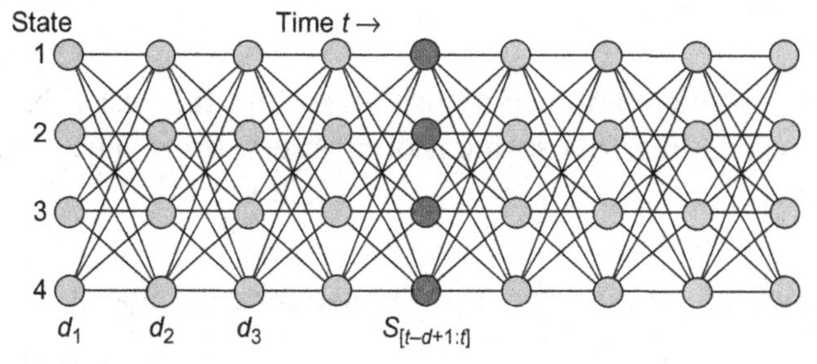

(a) A cut-set of all the paths.

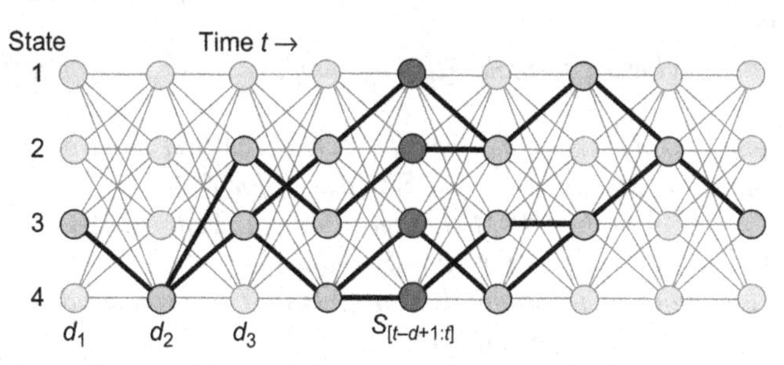

(b) Two trees consisting of the best subpaths.

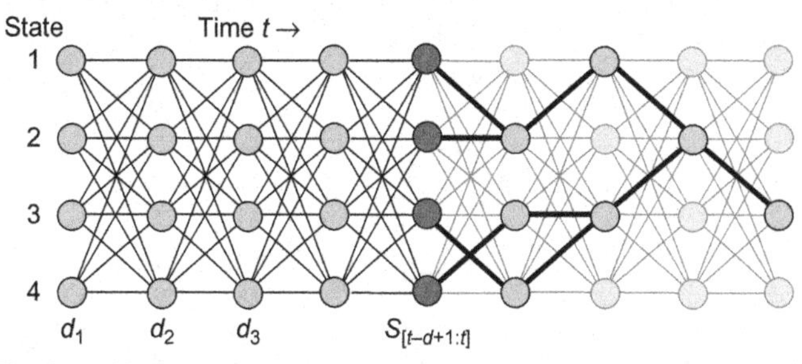

(c) The filtered best subpaths.

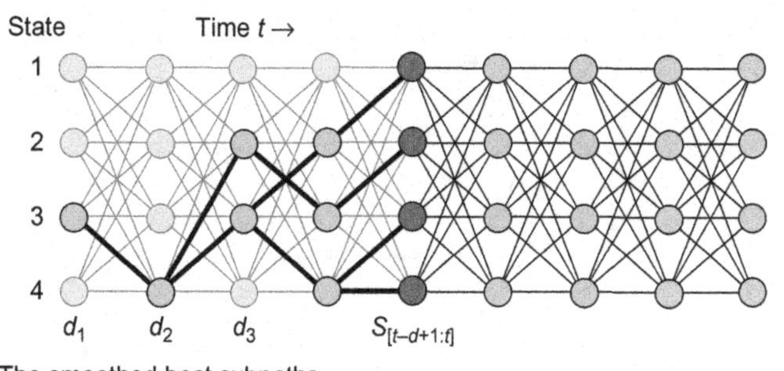

(d) The smoothed best subpaths.

Figure 2.2 The cut set of paths.
Each path from left to right in the trellis is an available state path. Each state in a state path has a dwell time. Suppose the critical time t *is given, and all states ending at* t *are considered. (a) For given observation sequence* $o_{1:T}$, *all possible state paths that have a state ending at given time* t. *(b) For given observation sequence* $o_{1:T}$, *all the best state paths that have a state ending at given time* t *are shown in the black lines, which form two opposite shortest trees. (c) If the previous states are not of concern, the future best state paths that the second state start at time* t+1 *are shown by the black lines, which form the shortest tree. (d) Conversely, if the future states are not of concern, the best paths that the last state end at time* t *are represented by the black lines, which form the shortest tree.*

We can use the combined algorithm to find the best subsequence of states corresponding to a segment of k observations. The algorithm for finding $\max_{S_{t+1:t+k}} P[S_{t+1:t+k}|o_{1:T}, \lambda]$ is as follows:

Algorithm 2.5 Finding the Best Subsequence of States

1. *For given t and k.*
2. *Use the forward formula* (2.6) *to compute $\alpha_t(j,d)$, for all $j \in \mathbf{S}$ and $d \in \mathbf{D}$.*
3. *Use the backward formula* (2.7) *to compute $\beta_{t+k}(j,d)$, for all $j \in \mathbf{S}$ and $d \in \mathbf{D}$.*
4. *Assume $\delta_{t'}(j,d) = \alpha_{t'}(j,d)$, for all $t' \leq t$. If the first state of the subpath is assumed to start at $t+1$, let $\delta_{t'}(j,d) = 0$ for $t' < t$.*
5. *Apply the Viterbi HSMM algorithm of* Eqn (2.29) *to compute $\delta_{t+1}(j,d), \ldots, \delta_{t+k}(j,d)$, for all $j \in \mathbf{S}$ and $d \in \mathbf{D}$.*
6. *The last state of the subpath that ends at $t+k$ can be determined by $(j_1^*, d_1^*) = \arg\max_{(j,d)} \delta_{t+k}(j,d)\beta_{t+k}(j,d)$.*
7. *Trace back following the similar procedure of the Viterbi HSMM algorithm. Then the best subpath of states, $S_{t+1:t+k}^*$, can be determined.*

CHAPTER 3

Parameter Estimation of General HSMM

This chapter discusses the maximum-likelihood estimation of model parameters for the general HSMM, including unsupervised, supervised, semi-supervised and online learning algorithms.

3.1 EM ALGORITHM AND MAXIMUM-LIKELIHOOD ESTIMATION

In the preceding problems, such as the forward–backward algorithm, MAP and maximum-likelihood (ML) estimation of states, it is assumed that the set of model parameters λ is given. If λ is unknown, it needs to learn about λ from the observations $o_{1:T}$: λ is initially estimated and then re-estimated so that the likelihood function $L(\lambda) = P[o_{1:T}|\lambda]$ increases up to its maximum value. Such training and updating process is referred to as *parameter re-estimation.*

3.1.1 EM Algorithm

The re-estimation procedure of the model parameters for HSMM is based on the following theorem.

Theorem 3.1

The parameter re-estimation procedure for the HSMMs increases the likelihood function of the model parameters.

We will prove this theorem using the expectation-maximization (EM) algorithm (Dempster et al., 1977). Let us consider *a posteriori probabilities* of the state sequence variable $S_{1:T} = (S_1, \ldots, S_T) \in \mathbf{S} \times \mathbf{S} \cdots \times \mathbf{S}$, given an instance of the observation sequence $o_{1:T}$, under the set of model parameters λ and its improved version λ', respectively. Then the *Kullback–Leibler divergence* (or *cross entropy*) between the two sets of model parameters is given by

$$\begin{aligned} D(\lambda||\lambda') &= \sum_{S_{1:T}} P[S_{1:T}|o_{1:T}, \lambda] \log\left(\frac{P[S_{1:T}|o_{1:T}, \lambda]}{P[S_{1:T}|o_{1:T}, \lambda']}\right) \\ &= \log\frac{L(\lambda')}{L(\lambda)} + \sum_{S_{1:T}} \frac{P[S_{1:T}, o_{1:T}|\lambda]}{L(\lambda)} \log\frac{P[S_{1:T}, o_{1:T}|\lambda]}{P[S_{1:T}, o_{1:T}|\lambda']}, \end{aligned} \tag{3.1}$$

where $L(\lambda) \equiv P[o_{1:T}|\lambda]$ is the *likelihood function* of the model parameters λ.

Following the discussion given by Ferguson (1980), an auxiliary function is defined as a conditional *expectation* (McLachlan and Krishnan, 2008)

$$\begin{aligned} Q(\lambda, \lambda') &\equiv E[\log P[S_{1:T}, o_{1:T}|\lambda']|o_{1:T}, \lambda] \\ &= \sum_{S_{1:T}} P[S_{1:T}, o_{1:T}|\lambda] \log P[S_{1:T}, o_{1:T}|\lambda'], \end{aligned} \tag{3.2}$$

where the sum is over all paths $S_{1:T}$, the first term $P[S_{1:T}, o_{1:T}|\lambda]$ is the probability of the path $S_{1:T}$ and observation sequence $o_{1:T}$ given the model parameter set λ, and the second term $P[S_{1:T}, o_{1:T}|\lambda']$ represents the probability of the same path and observation sequence given a second model parameter set λ'. Obviously, $Q(\lambda, \lambda) = \log \prod_{S_{1:T}} P[S_{1:T}, o_{1:T}|\lambda]^{P[S_{1:T}, o_{1:T}|\lambda]}$ is the log-likelihood function for all paths, and $Q(\lambda, \lambda') - Q(\lambda, \lambda)$ is the increment of the log-likelihood function brought by the new parameter set λ'.

Substituting Eqn (3.2) in Eqn (3.1), and considering that the Kullback–Leibler divergence is always nonnegative (Kullback and Leibler, 1951), that is, $D(\lambda||\lambda') \geq 0$, we have

$$\frac{Q(\lambda, \lambda') - Q(\lambda, \lambda)}{L(\lambda)} \leq \log \frac{L(\lambda')}{L(\lambda)}.$$

That is, if $Q(\lambda, \lambda') > Q(\lambda, \lambda)$, then $L(\lambda') > L(\lambda)$. In other words, by increasing the auxiliary function, the likelihood function can be increased. This implies that the best choice of λ' (for given λ) is found by solving the following *maximization* problem:

$$\max_{\lambda'} \frac{Q(\lambda, \lambda') - Q(\lambda, \lambda)}{L(\lambda)} = \max_{\lambda'} \sum_{S_{1:T}} \frac{P[S_{1:T}, o_{1:T}|\lambda]}{P[o_{1:T}|\lambda]} \log \frac{P[S_{1:T}, o_{1:T}|\lambda']}{P[S_{1:T}, o_{1:T}|\lambda]}. \tag{3.3}$$

Therefore, by iterating the *expectation step* (E-step) and the *maximization step* (M-step), as in the EM algorithm, an optimum λ' can be found.

3.1.2 Derivation of Re-estimation Formulas

Based on the EM algorithm, we can derive the ML estimation formulas of the model parameters. The state sequence $S_{1:T}$ underlying the given observation sequence $o_{1:T}$ is a series of random variables. Let $S_{1:T} = (i_1, d_1)\ldots(i_N, d_N)$, satisfying $\sum_{n=1}^{N} d_n = T$, denote a N-segment path, where (i_n, d_n) is a pair of random variables representing the nth state $i_n \in \mathbf{S}$ having duration $d_n \in \mathbf{D}$, and N is also a random variable. Given the path $S_{1:T} = (i_1, d_1)\ldots(i_N, d_N)$, the observation string is decomposed into segments $o_{t_0+1:t_1}, \ldots, o_{t_{N-1}+1:t_N}$, where $t_0 = 0$, $t_1 = d_1$, $t_2 = d_1 + d_2, \ldots, t_N = d_1 + \ldots + d_N = T$.

E-Step: Using this notation, we can express various smoothed probabilities, which can be used to express various expectations of model parameters.

The probability that state j having duration d and ending at time t is

$$\begin{aligned}
\eta_t(j,d) &= P[S_{[t-d+1:t]} = j, o_{1:T}|\lambda] \\
&= \sum_{S_{1:t-d}, S_{t+1:T}} P[S_{1:t-d}, S_{[t-d+1:t]} = j, S_{t+1:T}, o_{1:T}|\lambda] \\
&= \sum_{\substack{S_{1:T} \\ S_{[t-d+1:t]} = j}} P[S_{1:T}, o_{1:T}|\lambda] \\
&= \sum_N \sum_{n=1}^{N} \sum_{\substack{(i_m^N, d_m^N)_{m=1:N} \\ (i_n^N, d_n^N) = (j,d) \\ t_n = t}} P[(i_m^N, d_m^N)_{m=1:N}, o_{1:T}|\lambda] \\
&= \sum_N \sum_{n=1}^{N} \sum_{\substack{(i_n^N, d_n^N) = (j,\ d) \\ t_n = t}} P[(i_n^N, d_n^N), o_{1:T}|\lambda],
\end{aligned}$$

which can be used to compute the expectation of $b_{j,d}(v_{k_1:k_d})$ by letting

$$\hat{b}_{j,d}(v_{k_1:k_d}) = \sum_{t=d}^{T} \frac{\eta_t(j,d)}{P[o_{1:T}|\lambda]} \cdot I(o_{t-d+1:t} = v_{k_1:k_d}) \tag{3.4}$$

If let state j having duration d be the first state, then $t_1 = d_1 = d$, and

$$\begin{aligned}
\eta_d(j,d) &= P[S_{[1,d]} = j, o_{1:T}|\lambda] \\
&= \sum_N \sum_{(i_1^N, d_1^N) = (j,d)} P[(i_1^N, d_1^N), o_{1:T}|\lambda],
\end{aligned}$$

which can be used to compute the expectation of $\Pi_{j,d}$ by letting

$$\hat{\Pi}_{j,d} = \frac{\eta_d(j,d)}{P[o_{1:T}|\lambda]} \tag{3.5}$$

The probability that state i having duration h transits to state j having duration d is

$$\begin{aligned}\xi_t(i,h;j,d) &= P[S_{[t-h+1:t]}=i, S_{[t+1:t+d]}=j, o_{1:T}|\lambda]\\ &= \sum_{S_{1:t-h}, S_{t+d+1:T}} P[S_{1:t-h}, S_{[t-h+1:t]}=i, S_{[t+1:t+d]}=j, S_{t+d+1:T}, o_{1:T}|\lambda]\\ &= \sum_{\substack{S_{1:T}\\ S_{[t-h+1:t]}=i\\ S_{[t+1:t+d]}=j}} P[S_{1:T}, o_{1:T}|\lambda].\end{aligned}$$

If $S_{[t-h+1:t]}=i$ is the $(n-1)$th segment with ending time $t_{n-1}=t$, and $S_{[t+1:t+d]}=j$ is the nth segment, for $n=2,\ldots,N$, then

$$\begin{aligned}\xi_t(i,h;j,d) &= \sum_N \sum_{n=2}^{N} \sum_{\substack{(i_m^N, d_m^N)_{m=1:N}\\ (i_{n-1}^N, d_{n-1}^N)=(i,h)\\ (i_n^N, d_n^N)=(j,d)\\ t_{n-1}=t}} P[(i_m^N, d_m^N)_{m=1:N}, o_{1:T}|\lambda]\\ &= \sum_N \sum_{n=2}^{N} \sum_{\substack{(i_{n-1}^N, d_{n-1}^N)=(i,h)\\ (i_n^N, d_n^N)=(j,d)\\ t_{n-1}=t}} P[(i_{n-1}^N, d_{n-1}^N),(i_n^N, d_n^N), o_{1:T}|\lambda],\end{aligned}$$

which can be used to compute the expectation of $a_{(i,h)(j,d)}$ by letting

$$\hat{a}_{(i,h)(j,d)} = \sum_{t=h}^{T-d} \frac{\xi_t(i,h;j,d)}{P[o_{1:T}|\lambda]} \tag{3.6}$$

M-Step: After obtained the smoothed probabilities and expectations of the model parameters, we can update the model parameters by maximizing the likelihood function. Let (Ferguson, 1980)

$$\begin{aligned}P[(i_n^N, d_n^N)_{n=1:N}, o_{1:T}|\lambda] = P[(i_1^N, d_1^N)|\lambda]\\ \cdot \prod_{n=2}^{N} P[(i_n^N, d_n^N)|(i_{n-1}^N, d_{n-1}^N), \lambda]\\ \cdot \prod_{n=1}^{N} P[o_{t_{n-1}+1:t_n}|(i_n^N, d_n^N), \lambda],\end{aligned} \tag{3.7}$$

where (i_n^N, d_n^N) denotes the nth segment of the N-segment path $S_{1:T}=(i_1^N, d_1^N)\ldots(i_N^N, d_N^N)$. Now we try to derive the estimation formulas for the initial state distribution from $P[(i_1^N, d_1^N)|\lambda]$, the state transition probabilities from $\prod_{n=2}^{N} P[(i_n^N, d_n^N)|(i_{n-1}^N, d_{n-1}^N), \lambda]$, and the observation distributions from $\prod_{n=1}^{N} P[o_{t_{n-1}+1:t_n}|(i_n^N, d_n^N), \lambda]$.

Substituting Eqn (3.7) in Eqn (3.3), we get

$$\max_{\lambda'} \frac{Q(\lambda, \lambda') - Q(\lambda, \lambda)}{L(\lambda)} = \max_{\lambda'}(X_\pi(\lambda') + X_A(\lambda') + X_B(\lambda')),$$

where

$$X_\pi(\lambda') = \sum_{N} \sum_{(i_n^N, d_n^N)_{n=1:N}} \frac{P[(i_n^N, d_n^N)_{n=1:N}, o_{1:T}|\lambda]}{P[o_{1:T}|\lambda]} \log \frac{P[(i_1^N, d_1^N)|\lambda']}{P[(i_1^N, d_1^N)|\lambda]}$$

is related to the initial state distribution,

$$X_A(\lambda') = \sum_{N} \sum_{(i_n^N, d_n^N)_{n=1:N}} \frac{P[(i_n^N, d_n^N)_{n=1:N}, o_{1:T}|\lambda]}{P[o_{1:T}|\lambda]}$$
$$\log \prod_{n=2}^{N} \frac{P[(i_n^N, d_n^N)|(i_{n-1}^N, d_{n-1}^N), \lambda']}{P[(i_n^N, d_n^N)|(i_{n-1}^N, d_{n-1}^N), \lambda]}$$

is related to the state transition probabilities, and

$$X_B(\lambda') = \sum_{N} \sum_{(i_n^N, d_n^N)_{n=1:N}} \frac{P[(i_n^N, d_n^N)_{n=1:N}, o_{1:T}|\lambda]}{P[o_{1:T}|\lambda]}$$
$$\log \prod_{n=1}^{N} \frac{P[o_{t_{n-1}+1:t_n}|(i_n^N, d_n^N), \lambda']}{P[o_{t_{n-1}+1:t_n}|(i_n^N, d_n^N), \lambda]}$$

is regard the observation distributions. For the first term, we have

$$\begin{aligned}
X_\pi(\lambda') &= \sum_{N} \sum_{(i_1^N, d_1^N)} \frac{P[(i_1^N, d_1^N), o_{1:T}|\lambda]}{P[o_{1:T}|\lambda]} \log \frac{P[(i_1^N, d_1^N)|\lambda']}{P[(i_1^N, d_1^N)|\lambda]} \\
&= \sum_{(j,d)} \sum_{N} \sum_{(i_1^N, d_1^N) = (j,d)} \frac{P[(i_1^N, d_1^N), o_{1:T}|\lambda]}{P[o_{1:T}|\lambda]} \log \frac{P[(i_1^N, d_1^N)|\lambda']}{P[(i_1^N, d_1^N)|\lambda]} \\
&= \sum_{(j,d)} \log \frac{\Pi'_{j,d}}{\Pi_{j,d}} \sum_{N} \sum_{(i_1^N, d_1^N) = (j,d)} \frac{P[(i_1^N, d_1^N), o_{1:T}|\lambda]}{P[o_{1:T}|\lambda]} \\
&= \sum_{(j,d)} \log \frac{\Pi'_{j,d}}{\Pi_{j,d}} \frac{\eta_d(j,d)}{P[o_{1:T}|\lambda]},
\end{aligned}$$

where $\Pi_{j,d} \in \lambda$, $\Pi'_{j,d} \in \lambda'$, and $\Pi_{j,d} = P[S_{[1:d]} = j|\lambda] = \sum_{(i,h)} \pi_{i,h} a_{(i,h)(j,d)}$ is the distribution of the first states. For the second term, we have

$$X_A(\lambda') = \sum_N \sum_{n=2}^{N} \sum_{\substack{(i_{n-1}^N, d_{n-1}^N) \\ (i_n^N, d_n^N)}} \frac{P[(i_{n-1}^N, d_{n-1}^N), (i_n^N, d_n^N), o_{1:T}|\lambda]}{P[o_{1:T}|\lambda]}$$

$$\log \frac{P[(i_n^N, d_n^N)|(i_{n-1}^N, d_{n-1}^N), \lambda']}{P[(i_n^N, d_n^N)|(i_{n-1}^N, d_{n-1}^N), \lambda]}.$$

By combining like terms according to $(i, h), (j, d)$ and $i \neq j$, we have $P[(i_n^N, d_n^N)|(i_{n-1}^N, d_{n-1}^N), \lambda] = a_{(i,h)(j,d)}$, and

$$X_A(\lambda') = \sum_{\substack{(i,h),(j,d) \\ i \neq j}} \log \frac{a'_{(i,h)(j,d)}}{a_{(i,h)(j,d)}} \sum_N \sum_{n=2}^{N}$$

$$\sum_{\substack{(i_{n-1}^N, d_{n-1}^N) = (i,h) \\ (i_n^N, d_n^N) = (j,d)}} \frac{P[(i_{n-1}^N, d_{n-1}^N), (i_n^N, d_n^N), o_{1:T}|\lambda]}{P[o_{1:T}|\lambda]},$$

where $a_{(i,h)(j,d)} \in \lambda$ and $a'_{(i,h)(j,d)} \in \lambda'$. Further, by combining like terms according to the ending time of the state (i_{n-1}^N, d_{n-1}^N), it becomes

$$X_A(\lambda') = \sum_{\substack{(i,h),(j,d) \\ i \neq j}} \log \frac{a'_{(i,h)(j,d)}}{a_{(i,h)(j,d)}} \sum_{t=h}^{T-d} \sum_N \sum_{n=2}^{N}$$

$$\sum_{\substack{(i_{n-1}^N, d_{n-1}^N) = (i,h) \\ (i_n^N, d_n^N) = (j,d) \\ t_{n-1} = t}} \frac{P[(i_{n-1}^N, d_{n-1}^N), (i_n^N, d_n^N), o_{1:T}|\lambda]}{P[o_{1:T}|\lambda]}$$

$$= \sum_{\substack{(i,h),(j,d) \\ i \neq j}} \log \frac{a'_{(i,h)(j,d)}}{a_{(i,h)(j,d)}} \sum_{t=h}^{T-d} \frac{\xi_t(i,h;j,d)}{P[o_{1:T}|\lambda]},$$

For the third term, we have

$$X_B(\lambda') = \sum_N \sum_{n=1}^{N} \sum_{(i_n^N, d_n^N)} \frac{P[(i_n^N, d_n^N), o_{1:T}|\lambda]}{P[o_{1:T}|\lambda]} \log \frac{P[o_{t_{n-1}+1:t_n}|(i_n^N, d_n^N), \lambda']}{P[o_{t_{n-1}+1:t_n}|(i_n^N, d_n^N), \lambda]}.$$

By combining like terms according to (j, d) and observable substring $v_{k_1:k_d}$, we have $P[o_{t_{n-1}+1:t_n}|(i_n^N, d_n^N), \lambda] = b_{j,d}(v_{k_1:k_d})$, and

$$X_B(\lambda') = \sum_{(j,d)} \sum_{v_{k_1:k_d}} \log \frac{b'_{j,d}(v_{k_1:k_d})}{b_{j,d}(v_{k_1:k_d})} \sum_N \sum_{n=1}^{N} \sum_{\substack{(i_n^N, d_n^N) = (j,d) \\ o_{t_{n-1}+1:t_n} = v_{k_1:k_d}}} \frac{P[(i_n^N, d_n^N), o_{1:T}|\lambda]}{P[o_{1:T}|\lambda]},$$

where $b_{j,d}(v_{k_1:k_d}) \in \lambda$, $b'_{j,d}(v_{k_1:k_d}) \in \lambda'$. Further, by combining like terms according to the ending time of the state,

$$X_B(\lambda') = \sum_{(j,d)} \sum_{v_{k_1:k_d}} \log \frac{b'_{j,d}(v_{k_1:k_d})}{b_{j,d}(v_{k_1:k_d})} \sum_{t=d}^{T} \sum_N \sum_{n=1}^{N} \sum_{\substack{(i_n^N, d_n^N) = (j,d) \\ o_{t_{n-1}+1:t_n} = v_{k_1:k_d} \\ t_n = t}} \frac{P[(i_n^N, d_n^N), o_{1:T}|\lambda]}{P[o_{1:T}|\lambda]}$$

$$= \sum_{(j,d)} \sum_{v_{k_1:k_d}} \log \frac{b'_{j,d}(v_{k_1:k_d})}{b_{j,d}(v_{k_1:k_d})} \sum_{t=d}^{T} \frac{\eta_t(j,d)}{P[o_{1:T}|\lambda]} \cdot \mathrm{I}(o_{t-d+1:t} = v_{k_1:k_d}),$$

where the indicator function $\mathrm{I}(o_{t-d+1:t} = v_{k_1:k_d}) = 1$ if $v_{k_1} = o_{t-d+1}, \ldots, v_{k_d} = o_t$ and 0 otherwise.

Then Eqn (3.3) becomes

$$\max_{\lambda'} \frac{Q(\lambda, \lambda') - Q(\lambda, \lambda)}{L(\lambda)} = \max_{\lambda'} \left\{ \sum_{j,d} \hat{\Pi}_{j,d} \log \frac{\Pi'_{j,d}}{\Pi_{j,d}} + \sum_{\substack{(i,h),(j,d) \\ i \neq j}} \hat{a}_{(i,h)(j,d)} \log \frac{a'_{(i,h)(j,d)}}{a_{(i,h)(j,d)}} + \sum_{j,d} \sum_{v_{k_1:k_d}} \hat{b}_{j,d}(v_{k_1:k_d}) \log \frac{b'_{j,d}(v_{k_1:k_d})}{b_{j,d}(v_{k_1:k_d})} \right\},$$

where $\lambda' = \{\Pi'_{j,d}, a'_{(i,h)(j,d)}, b'_{j,d}(v_{k_1:k_d})\}$ are parameters to be found by maximizing the auxiliary function, and $\{\hat{\Pi}_{j,d}, \hat{a}_{(i,h)(j,d)}, \hat{b}_{j,d}(v_{k_1:k_d})\}$ are expectations for the given set of model parameters λ, which have been determined in the E-step.

It can be proved that when we choose the estimated values by letting $\Pi'_{j,d} = \hat{\Pi}_{j,d}$, $a'_{(i,h)(j,d)} = \hat{a}_{(i,h)(j,d)}$, and $b'_{j,d}(v_{k_1:k_d}) = \hat{b}_{j,d}(v_{k_1:k_d})$, s.t., $\sum_{(j,d)} \Pi'_{j,d} = 1$, $\sum_{j \neq i,d} a'_{(i,h)(j,d)} = 1$, and $\sum_{v_{k_1:k_d}} b'_{j,d}(v_{k_1:k_d}) = 1$, the auxiliary function and the likelihood function are maximized.

Obviously, if $\Pi'_{j,d}$, $a'_{(i,h)(j,d)}$ and $b'_{j,d}(v_{k_1:k_d})$ are parametric distributions with parameters ρ, θ and φ, respectively, the auxiliary function and the likelihood function can be maximized by

$$\max_{\rho} X_{\pi}(\rho) = \max_{\rho} \sum_{(j,d)} \frac{\eta_d(j,d)}{P[o_{1:T}|\lambda]} \log \frac{\Pi'_{j,d}(\rho)}{\Pi_{j,d}}, \tag{3.8}$$

$$\max_{\theta} X_A(\theta) = \max_{\theta} \sum_{\substack{(i,h),(j,d) \\ i \neq j}} \left(\sum_{t=h}^{T-d} \frac{\xi_t(i,h;j,d)}{P[o_{1:T}|\lambda]} \right) \log \frac{a'_{(i,h)(j,d)}(\theta)}{a_{(i,h)(j,d)}} \tag{3.9}$$

and

$$\max_{\varphi} X_B(\varphi) = \max_{\varphi} \sum_{(j,d)} \sum_{t=d}^{T} \frac{\eta_t(j,d)}{P[o_{1:T}|\lambda]} \log \frac{b'_{j,d}(o_{t-d+1:t}, \varphi)}{b_{j,d}(o_{t-d+1:t})}. \tag{3.10}$$

Instead of using the forward–backward algorithms for the E-step and the re-estimation formulas for the M-step, Krishnamurthy and Moore (1991) use the EM algorithm to directly re-estimation the model parameters. In this case, the unknowns are the model parameters λ' as well as the state sequence $S_{1:T}$. MCMC sampling is a general methodology for generating samples from a desired probability distribution function and the obtained samples are used for various types of inference. Therefore, MCMC sampling can also be used in the estimation of the state sequence and the model parameters (Djuric and Chun, 1999, 2002). MCMC sampling draws samples of the unknowns from their posteriors so that the posteriors can be approximated using these samples. The prior distributions of all the unknowns are required to be specified before applying the MCMC sampling methods. The emission probabilities and the durations of the various states are often modeled using some parametric distributions.

By assuming the HSMM is a variable duration HMM, that is, observations are conditionally independent of each other and self-transition of a state is allowed and independent of the previous state, the model can be expressed using an HMM. Based on the HMM theory, Barbu and Limnios (2008) investigated the asymptotic properties of the nonparametric ML estimator, and proved its consistency and asymptotic normality.

3.2 RE-ESTIMATION ALGORITHMS OF MODEL PARAMETERS

From the EM algorithm we can see that the expectations of parameters can be used for estimating the model parameters, which can be calculated by the forward–backward variables.

3.2.1 Re-Estimation Algorithm for the General HSMM

The model parameters, $\hat{b}_{j,d}(v_{k_1:k_d})$, $\hat{\Pi}_{j,d}$, and $\hat{a}_{(i,h)(j,d)}$, for the general HSMM can be estimated by Eqns (3.4)–(3.6), respectively, where $\hat{a}_{(i,h)(i,h)} = 0$, for all i, and the model parameters must be normalized such that $\sum_{j,d} \hat{\Pi}_{j,d} = 1$, $\sum_{j,d} \hat{a}_{(i,h)(j,d)} = 1$, and $\sum_{v_{k_1},\ldots,v_{k_d}} b_{j,d}(v_{k_1:k_d}) = 1$, for $i, j \in \mathbf{S}$, $h, d \in \mathbf{D}$, and $v_{k_1:k_d} \in \mathbf{V}^d$.

Except those parameters for the general HSMM, parameters for other variants of HSMMs can be estimated as well, such as, the state transition probabilities

$$\hat{a}_{ij} = \sum_{t=1}^{T-1} \xi_t(i,j) / \sum_{j \neq i} \sum_{\tau=1}^{T-1} \xi_\tau(i,j), \tag{3.11}$$

the duration probabilities of state j

$$\hat{p}_j(d) = \sum_{t=d}^{T} \eta_t(j,d) / \sum_{h} \sum_{\tau=h}^{T} \eta_\tau(j,h), \tag{3.12}$$

the observation probabilities for given state j

$$\hat{b}_j(v_k) = \sum_{t=1}^{T} [\gamma_t(j) \cdot \mathrm{I}(o_t = v_k)] / \sum_{\tau=1}^{T} \gamma_\tau(j), \tag{3.13}$$

and the initial state distribution

$$\hat{\pi}_j = \gamma_0(j) / \sum_j \gamma_0(j), \tag{3.14}$$

subject to $\sum_{j \neq i} \hat{a}_{ij} = 1$, $\sum_d \hat{p}_j(d) = 1$, and $\sum_{v_k} b_j(v_k) = 1$.

The parameter re-estimation procedure must be implemented iteratively until the likelihood increases to the maximum or a fixed point.

Algorithm 3.1 Re-estimation Algorithm

1. *Assume an initial model parameter set* λ_0, *and let* $l = 0$.
2. *For given model parameter set* λ_l, *use the forward–backward formulas of* Eqn (2.6) *and* Eqn (2.7) *to compute the forward and backward variables* $\{\alpha_t(j,d)\}$ *and* $\{\beta_t(j,d)\}$.
3. *Use the forward and backward variables to compute the related probabilities* $\eta_t(j,d), \xi_t(i,h;j,d), \xi_t(i,j)$ *and* $\gamma_t(j)$ *by* Eqn (2.10) *through* Eqn (2.14).
4. *Estimate the model parameters using* Eqns (3.4)–(3.6) *to get* $\hat{\lambda}_{l+1}$.
5. *Let* $\lambda_{l+1} = \hat{\lambda}_{l+1}$ *and* $l++$.
6. *Repeat step 2 to step 5 until the likelihood* $L(\lambda_l) = P[o_{1:T}|\lambda_l]$ *increases to the maximum or a fixed point.*

3.2.2 Supervised/Semi-Supervised Learning

In some applications, both the observation sequence and state sequence are given, but the model parameters are required to be estimated. For example, in digital communication area, when a given bit stream is coded and sent to test the channel property, the state sequence is known and the observation sequence can be obtained. Then the model parameters that are used to describe the channel property are to be estimated based on both the state sequence and observation sequence.

For the general cases, we suppose a part of states in the state sequence and their durations $S_{[t_1-d_1+1:t_1]} = j_1, \ldots, S_{[t_n-d_n+1:t_n]} = j_n$ are known in advance and others are hidden/unknown for a given observation sequence $o_{1:T}$, where the time intervals $[t_1 - d_1 + 1 : t_1], \ldots, [t_n - d_n + 1 : t_n]$ are not overlapped and not necessarily consecutive, that is, $t_{k-1} \leq t_k - d_k$, for $k = 2, \ldots, n$. In this case, the semi-supervised learning can be implemented. For a known state $S_{[t_k-d_k+1:t_k]} = j_k$, the other states that overlap with it are all invalid, that is, $S_{[t-d+1:t]} = j$, for $(j,d,t) \neq (j_k, d_k, t_k)$ and

$[t-d+1,t] \cap [t_k-d_k+1,t_k] \neq \phi$, are invalid states. Therefore, in the model parameter estimation procedure, it must keep the forward and backward variables corresponding to the known states to be nonzeros and the invalid states zeros. Define $\mathbf{L} = \{(j_k, d_k, t_k), k = 1, \ldots, n\}$ as the set of labeled/known states and $\mathbf{U} = [t_1 - d_1 + 1, t_1] \cup \ldots \cup [t_n - d_n + 1, t_n]$ as the time periods that the states are labeled. Then let $\alpha_t(j,d) = 0$ and $\beta_t(j,d) = 0$ if $(j,d,t) \notin \mathbf{L}$ but $[t-d+1,t] \cap \mathbf{U} \neq null$.

Applying those constraints in the re-estimation procedure of the model parameters, the model parameters can then be learned in a semi-supervised manner. That is, the forward formula (2.6) becomes

$$\alpha_t(j,d) = \{1 - \mathrm{I}((j,d,t) \notin \mathbf{L}) \cdot \mathrm{I}([t-d+1,t] \cap \mathbf{U} \neq \phi)\} \cdot \sum_{i \neq j;h} \alpha_{t-d}(i,h) \cdot a_{(i,h)(j,d)} \cdot b_{j,d}(o_{t-d+1:t})$$

and the backward formula (2.7) can be similarly changed. After all the forward and backward variables are obtained, the model parameters can be estimated using Algorithm 3.1.

Obviously, if $t_1 = d_1$ and $d_1 + \ldots + d_n = T$, then the entire state sequence $s_{1:T}$ is labeled, and the model parameters can be learned in a supervised learning manner. In this case,

$$\alpha_{t_k}(j_k, d_k) = \alpha_{t_{k-1}}(j_{k-1}, d_{k-1}) \cdot a_{(j_{k-1},d_{k-1})(j_k,d_k)} \cdot b_{j_k,d_k}(o_{t_k-d_k+1:t_k})$$

for $k = 2, \ldots, n$, where $\alpha_{t_1}(j_1, d_1) = \Pi_{j_1,d_1} \cdot b_{j_1,d_1}(o_{1:t_1})$. All other forward variables $\alpha_t(j,d)$ are zeros. The backward variables $\beta_t(j,d)$ can be similarly obtained. In this way, the model parameters can be learned using Algorithm 3.1.

3.2.3 Multiple Observation Sequences

In practice, it is often that there exist multiple observation sequences. The underlying state sequences may be different, but the model parameters are not changed. For example, network traffic may vary a lot because the underlying number of users is changing. If the arrival rate of packets at the same network node is collected in the same working hour of recent days, the multiple observation sequences collected are assumed to be governed by the same model parameters.

Now we derive the estimation formulas of the model parameters when there are multiple observation sequences. Let $\mathbf{o}^{(k)} = (o_1^{(k)}, o_2^{(k)}, \ldots, o_{T_k}^{(k)})$ denote the kth observation sequence, $\mathbf{S}^{(k)} = (S_1^{(k)}, S_2^{(k)}, \ldots, S_{T_k}^{(k)})$ the corresponding state sequence, and T_k the sequence length, $k = 1, \ldots, n$. Then the auxiliary function Eqn (3.2) becomes

$$
\begin{aligned}
Q(\lambda, \lambda') &= \sum_{\mathbf{S}^{(1)} \cdots \mathbf{S}^{(n)}} P[\mathbf{S}^{(1)} \cdots \mathbf{S}^{(n)}, \mathbf{o}^{(1)} \cdots \mathbf{o}^{(n)} | \lambda] \log P[\mathbf{S}^{(1)} \cdots \mathbf{S}^{(n)}, \mathbf{o}^{(1)} \cdots \mathbf{o}^{(n)} | \lambda'] \\
&= \sum_{\mathbf{S}^{(1)} \cdots \mathbf{S}^{(n)}} \prod_{k'=1}^{n} P[\mathbf{S}^{(k')}, \mathbf{o}^{(k')} | \lambda] \log \prod_{k=1}^{n} P[\mathbf{S}^{(k)}, \mathbf{o}^{(k)} | \lambda'] \\
&= \sum_{k=1}^{n} \sum_{\mathbf{S}^{(1)} \cdots \mathbf{S}^{(n)}} \prod_{k'=1}^{n} P[\mathbf{S}^{(k')}, \mathbf{o}^{(k')} | \lambda] \log P[\mathbf{S}^{(k)}, \mathbf{o}^{(k)} | \lambda'] \\
&= \sum_{k=1}^{n} \frac{L(\lambda)}{P[\mathbf{o}^{(k)} | \lambda]} \sum_{\mathbf{S}^{(k)}} P\left[\mathbf{S}^{(k)}, \mathbf{o}^{(k)} | \lambda\right] \log P\left[\mathbf{S}^{(k)}, \mathbf{o}^{(k)} | \lambda'\right].
\end{aligned}
$$

where the likelihood function $L(\lambda) = P[\mathbf{o}^{(1)}, \ldots, \mathbf{o}^{(n)} | \lambda] = \prod_{k'=1}^{n} P[\mathbf{o}^{(k')} | \lambda]$, and $L^{(k)}(\lambda) = P[\mathbf{o}^{(k)} | \lambda]$ is the likelihood function of the kth observation sequence. Therefore, by defining

$$
\frac{Q^{(k)}(\lambda, \lambda') - Q^{(k)}(\lambda, \lambda)}{L^{(k)}(\lambda)} = \sum_{\mathbf{S}^{(k)}} \frac{P[\mathbf{S}^{(k)}, \mathbf{o}^{(k)} | \lambda]}{P[\mathbf{o}^{(k)} | \lambda]} \log \frac{P[\mathbf{S}^{(k)}, \mathbf{o}^{(k)} | \lambda']}{P[\mathbf{S}^{(k)}, \mathbf{o}^{(k)} | \lambda]},
$$

the maximization problem of Eqn (3.3) becomes

$$
\max_{\lambda'} \frac{Q(\lambda, \lambda') - Q(\lambda, \lambda)}{L(\lambda)} = \max_{\lambda'} \sum_{k=1}^{n} \frac{Q^{(k)}(\lambda, \lambda') - Q^{(k)}(\lambda, \lambda)}{L^{(k)}(\lambda)}.
$$

Following the similar derivations of the EM algorithm, the expectations become

$$
\hat{\Pi}_{j,d} = \sum_{k=1}^{n} \frac{\eta_d^{(k)}(j, d)}{P[\mathbf{o}^{(k)} | \lambda]},
$$

$$
\hat{a}_{(i,h)(j,d)} = \sum_{k=1}^{n} \sum_{t=h}^{T_k - d} \frac{\xi_t^{(k)}(i, h; j, d)}{P[\mathbf{o}^{(k)} | \lambda]},
$$

and

$$
\hat{b}_{j,d}(v_{k_1:k_d}) = \sum_{k=1}^{n} \sum_{t=d}^{T_k} \frac{\eta_t^{(k)}(j, d)}{P[\mathbf{o}^{(k)} | \lambda]} \cdot \mathrm{I}(o_{t-d+1:t}^{(k)} = v_{k_1:k_d}),
$$

where $\xi_t^{(k)}(i,h;j,d)$ and $\eta_t^{(k)}(j,d)$ are the variables calculated using the kth observation sequence $\mathbf{o}^{(k)}$ for given set of model parameters λ. Again, the model parameters must be normalized so that $\sum_{j,d} \hat{\Pi}_{j,d} = 1$, $\sum_{j \neq i,d} \hat{a}_{(i,h)(j,d)} = 1$, and $\sum_{v_{k_1},\ldots,v_{k_d}} \hat{b}_{j,d}(v_{k_1:k_d}) = 1$. Besides,

$$\hat{a}_{ij} = \sum_{k=1}^{n} \sum_{t=1}^{T_k - 1} \frac{\xi_t^{(k)}(i,j)}{P[\mathbf{o}^{(k)}|\lambda]},$$

$$\hat{p}_j(d) = \sum_{k=1}^{n} \sum_{t=d}^{T_k} \frac{\eta_t^{(k)}(j,d)}{P[\mathbf{o}^{(k)}|\lambda]},$$

$$\hat{b}_j(v) = \sum_{k=1}^{n} \sum_{t=1}^{T_k} \left[\frac{\gamma_t^{(k)}(j)}{P[\mathbf{o}^{(k)}|\lambda]} \cdot I(o_t^{(k)} = v) \right],$$

and

$$\hat{\pi}_j = \sum_{k=1}^{n} \frac{\gamma_0^{(k)}(j)}{P[\mathbf{o}^{(k)}|\lambda]},$$

subject to $\sum_{j \neq i} \hat{a}_{ij} = 1, \sum_d \hat{p}_j(d) = 1$, and $\sum_{v_k} b_j(v_k) = 1$, where $\gamma_t^{(k)}(j)$ are corresponding to the kth observation sequence.

Algorithm 3.2 Re-estimation Algorithm for Multiple Observation Sequences

1. *Assume an initial model parameter set* λ_0, *and let* $l = 0$.
2. *For given model parameter set* λ_l, *use the forward–backward formulas of* Eqn (2.6) *and* Eqn (2.7) *to compute the forward and backward variables* $\{\alpha_t^{(k)}(j,d)\}$ *and* $\{\beta_t^{(k)}(j,d)\}$ *over the k'th observation sequence, for* $k = 1,\ldots,n$.
3. *Use the forward and backward variables to compute the related probabilities* $\eta_t^{(k)}(j,d)$, $\xi_t^{(k)}(i,h;j,d)$, $\xi_t^{(k)}(i,j)$ *and* $\gamma_t^{(k)}(j)$ *by* Eqn (2.10) *through* Eqn (2.14), *for* $k = 1,\ldots,n$.
4. *Estimate the model parameters using* $\hat{\Pi}_{j,d} = \sum_{k=1}^{n} \frac{\eta_d^{(k)}(j,d)}{P[\mathbf{o}^{(k)}|\lambda]}$, $\hat{a}_{(i,h)(j,d)} = \sum_{k=1}^{n} \sum_{t=h}^{T_k - d} \frac{\xi_t^{(k)}(i,h;j,d)}{P[\mathbf{o}^{(k)}|\lambda]}$, *and* $\hat{b}_{j,d}(v_{k_1:k_d}) = \sum_{k=1}^{n} \sum_{t=d}^{T_k} \frac{\eta_t^{(k)}(j,d)}{P[\mathbf{o}^{(k)}|\lambda]} \cdot \mathrm{I}(o_{t-d+1:t}^{(k)} = v_{k_1:k_d})$ *to get* $\hat{\lambda}_{l+1}$.
5. *Let* $\lambda_{l+1} = \hat{\lambda}_{l+1}$ *and* $l{+}{+}$.
6. *Repeat step 2 to step 5 until the likelihood* $L(\lambda_l) = P[\mathbf{o}^{(1)},\ldots,\mathbf{o}^{(n)}|\lambda_l] = \prod_{k=1}^{n} P[\mathbf{o}^{(k)}|\lambda_l]$ *increases to the maximum or a fixed point.*

3.3 ORDER ESTIMATION OF HSMM

In the re-estimation algorithms discussed so far, the number of hidden states, M, the maximum length of state duration, D, the number of observable values, K, and the length of the observation sequence, T, are usually assumed known in the context of applications. However, there exist the learning issues that the order M is unknown in practice. A detailed discussion on the order estimation of HMMs can be found in Ephraim and Merhav (2002)'s Section VIII, and the issues of over-fitting and model selection in Ghahramani (2001)'s Section 7. It can be seen that order estimation of HMMs is a difficult problem. There are no theoretically rigorous approaches to the problem of order estimation of HMMs available at the present time. Order estimation of HSMMs is even more difficult than that of HMMs, because HSMMs have variable durations. Therefore, for an HSMM we must estimate both the number of states, M, and the maximum length of state durations, D.

Some special HSMMs can be described by a DBN using a directed graphical model. For simplicity, one usually assumes the observations are conditionally independent, that is,

$$b_{j,d}(o_{t+1:t+d}) = P[o_{t+1:t+d} | S_{[t+1:t+d]} = j] = \prod_{\tau=t+1}^{t+d} b_j(o_\tau), \tag{3.15}$$

where $b_j(v_k) \equiv P[o_t = v_k | S_t = j]$. To easily identify when a segment of states starts, one usually further assumes a state transition is independent of the previous state duration. This is just the assumption made for the explicit duration HMM and the residual time HMM. The conditional probability distribution (CPD) function for the explicit duration HMM is (Murphy, 2002a)

$$P[S_t = j | S_{t-1} = i, R_{t-1} = \tau] = \begin{cases} a_{ij}, & \text{if } \tau = 1 (\text{transition}) \\ \mathrm{I}(i = j), & \text{if } \tau > 1 (\text{decrement}) \end{cases}$$

$$P[R_t = \tau' | S_t = j, R_{t-1} = \tau] = \begin{cases} p_j(\tau'), & \text{if } \tau = 1 (\text{transition}) \\ \mathrm{I}(\tau' = \tau - 1), & \text{if } \tau > 1 (\text{decrement}) \end{cases}$$

and for the residual time HMM is

$$P[Q_t = (j, \tau') | Q_{t-1} = (i, \tau)] = \begin{cases} a_{i(j,\tau')}, & \text{if } \tau = 1 (\text{transition}) \\ \mathrm{I}(\tau' = \tau - 1) \cdot I(i = j), & \text{if } \tau > 1 (\text{decrement}) \end{cases}$$

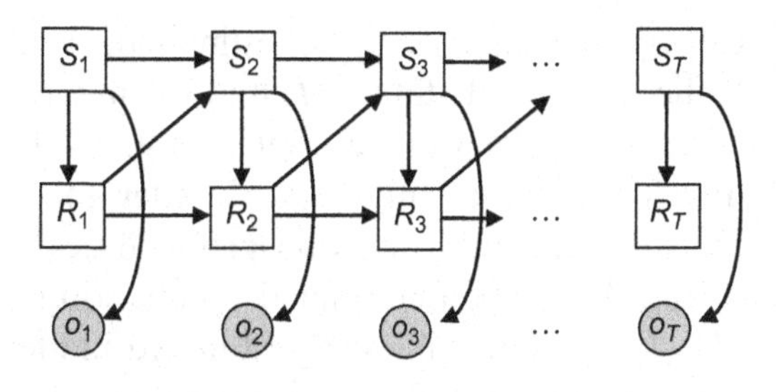

(a) DBN for explicit duration HMM

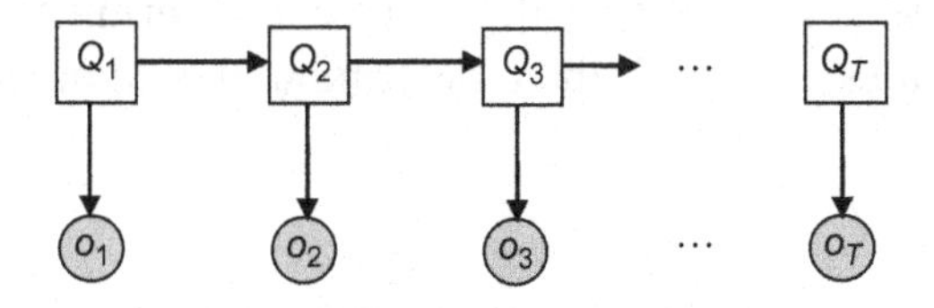

(b) DBN for residual time HMM

Figure 3.1 DBNs of the HSMMs.
A HSMM can be described by a DBN using a directed graphical model. In this figure, each slice has one or two discrete hidden nodes and one observed node. Clear means hidden, shaded means observed. The arc from node A to node B indicates that B "depends on" A. (a) R_t *is the remaining duration of state* S_t *and is dependent on* S_t*;* S_{t+1} *is dependent on both* R_t *and* S_t*, for* $t = 1, \ldots, T$. *(b)* $Q_t = (j,\tau)$ *is a complex state that represents* $S_t = j$ *and* $R_t = \tau$*;* Q_{t+1} *is dependent on* Q_t.

where R_t is the remaining duration of state S_t, $Q_t = (j, \tau)$ represents $S_t = j$ and $R_t = \tau$, and the self-transition probability $a_{i(i,\tau')} = 0$. The indicator function $\mathrm{I}(x) = 1$ if x is true and zero otherwise.

Figure 3.1 are directed graphical models for describing the HSMMs. Some other DBNs for HSMMs are presented in Murphy (2002a). In fact, we can alternatively use an undirected graphical model to describe the HSMMs and to learn the unknown quantities, such as semi-Markov conditional random fields (semi-CRFs) introduced by Sarawagi and Cohen (2004). In this model, the assumption that the observations are conditionally independent is not needed.

As discussed by Ghahramani (2001), a Bayesian approach for learning treats all unknown quantities as random variables. These unknown quantities comprise the number of states, the parameters, and the hidden states. By integrating over both the parameters and the hidden states, the unknown quantities can be estimated. For the explicit duration HMM, the number of states, M, and the maximum length of state durations, D, can be determined after S_t and R_t, for $t=1$ to T, are estimated. For the residual time HMM, after the set of hidden states that Q_t can take and the transition probabilities are estimated, the values of M and D can be determined by checking the transition probabilities of $P[q_t|q_{t-1}]$, where q_t is the estimated hidden state of Q_t. Obviously from the CPDs, $P[q_t|q_{t-1}]=1$ represents a self-transition, and $P[q_t|q_{t-1}]<1$ a state transition. Therefore, by counting the number of consecutive self-transitions we can determine the maximum duration of states, D, and then determine the number of HSMM states, M.

3.4 ONLINE UPDATE OF MODEL PARAMETERS

Re-estimation of the model parameters using the forward–backward algorithms has been discussed in the previous sections. However, these algorithms require the backward procedures and the iterative calculations, and so are not practical for online learning when the length T of the observation sequence is increasing. A few of online algorithms for HSMM have been developed in the literature, including an adaptive EM algorithm by Ford et al. (1993), an online algorithm based on recursive prediction error (RPE) techniques by Azimi et al. (2003, 2005), and a recursive ML estimation (RMLE) algorithm by Squire and Levinson (2005).

3.4.1 Online Update Using Forward-Only Algorithm

An online update algorithm for the model parameters can be based on the forward-only algorithm of Algorithm 2.3.

To update the observation/emission probabilities online, an expectation vector for the observation/emission probabilities is defined by $\sum_{\tau=1}^{T}\mathbf{I}_\tau\vec{\eta}_\tau=\sum_{\tau=1}^{T}\mathbf{I}_\tau Diag(\vec{\alpha}_\tau)\vec{\beta}_\tau$, where $\vec{\eta}_t=(\eta_t(1,1),\ldots,\eta_t(1,D),\ldots,\eta_t(M,1),\ldots,\eta_t(M,D))^T=Diag(\vec{\alpha}_t)\vec{\beta}_t$. Using the forward-only algorithm of Algorithm 2.3, the expectation vector $\vec{g}_T^{\,B}=\sum_{\tau=1}^{T}\mathbf{I}_\tau\vec{\eta}_\tau=\vec{G}\left(\sum_{\tau=1}^{T}\mathbf{I}_\tau Diag(\vec{\alpha}_\tau)\vec{\beta}_\tau\right)$ can be determined. Let

$k = MD(l-1) + (j-1)D + d$ and $\hat{b}_{j,d}(v_{d,l}) = \vec{g}_T^B(k)$, for $j = 1, \ldots, M$, $d = 1, \ldots, D$, and $l = 1, \ldots, L$. Then the observation/emission probability matrix can be updated at time T.

Because the length of the observation sequence is increasing, new observable values and substrings may be found. Suppose $v_{k_1:k_d}$ is a new substring that has never been observed. The current observation probability $b_{j,d}(v_{k_1:k_d})$ equals 0. Therefore, when $v_{k_1:k_d}$ is firstly observed, we must let $b_{j,d}(v_{k_1:k_d})$ equal a very small probability, say $0.5 \times \min_{v_{k'_1:k'_d}} \{b_{j,d}(v_{v_{k'_1:k'_d}}) > 0\}$; otherwise, at time T, we will get $\hat{b}_{j,d}(v_{k_1:k_d}) = 0$.

To update the state transition probabilities online, an expectation vector is defined by $\sum_{t=2}^{T} \sum_{d=1}^{\min\{D,t-1\}} \mathbf{Z}_{t-d} \mathbf{B}_{t-d+1} \mathbf{E}_d \vec{\beta}_t$. Using the forward-only algorithm of Algorithm 2.3, the expectation vector of the state transition probabilities $\vec{g}_T^A = \vec{G}\left(\sum_{t=2}^{T} \sum_{d=1}^{\min\{D,t-1\}} \mathbf{Z}_{t-d} \mathbf{B}_{t-d+1} \mathbf{E}_d \vec{\beta}_t\right)$ can be determined. Let $k = (i-1)MD^2 + (h-1)MD + (j-1)D + d$ and $\hat{a}_{(i,h)(j,d)} = \vec{g}_T^A(k)$, for $i, j \in \mathbf{S}$ and $h, d \in \mathbf{D}$. Then the state transition probability matrix can be updated online.

To update the probabilities of the starting states online, though this is not often required for the online update, an expectation vector is defined by $\sum_{\tau=1}^{T} \mathbf{Y}_\tau \vec{\beta}_\tau$ with $\mathbf{Y}_d = Diag(\vec{\alpha}_1^*)\mathbf{E}_d$, for $d = 1, \ldots, D$, and $\mathbf{Y}_\tau = 0$, for $\tau > D$. Then the expectation vector $\vec{g}_T^\Pi = \vec{G}\left(\sum_{\tau=1}^{T} \mathbf{Y}_\tau \vec{\beta}_\tau\right)$ can be determined. The probabilities $\hat{\Pi}_{j,d}$ of the starting states can be updated online by letting $\hat{\Pi}_{j,d} = \vec{g}_T^\Pi(j,d)$.

According to Theorem 3.1, using λ to compute the expectation vector and then get the new set of model parameters $\hat{\lambda}$ is a likelihood increasing procedure. This online algorithm is an adaptive updating procedure that fits to the slow varying model parameters.

Theorem 3.2

The forward-only algorithm based online update algorithm is a likelihood increasing and adaptive updating procedure. It is suitable for online updating the model parameters that do not change with time or vary slowly with time. If the HSMM process is time-homogeneous, the updating model parameters are expected to converge as the length of the observation sequence increases.

To prove Theorem 3.2, we denote $\mathbf{X}_t = (\mathbf{X}_t(1), \ldots, \mathbf{X}_t(D))$,

$$\mathbf{F} = \begin{bmatrix} \mathbf{0} & & \cdots & & \mathbf{0} \\ \mathbf{E} & \ddots & & & \\ \mathbf{0} & \ddots & \ddots & & \vdots \\ \vdots & \ddots & \ddots & \ddots & \\ \mathbf{0} & \cdots & \mathbf{0} & \mathbf{E} & \mathbf{0} \end{bmatrix}$$

and $\mathbf{W}_{t+1} = (\mathbf{AB}'_{t+1} + \mathbf{Y}_{t+1}, \mathbf{AB}'_{t+2}, \ldots, \mathbf{AB}'_{t+D})$ for simplicity, where $\mathbf{E}$ is the identity matrix and $\mathbf{F}^D = \mathbf{0}$. Then the forward-only formula (2.26) can be rewritten as

$$\mathbf{X}_{t+1} = \mathbf{X}_t\mathbf{F} + \mathbf{W}_{t+1}.$$

Therefore, for $T > D$,

$$\begin{aligned} \mathbf{X}_T &= \mathbf{X}_{T-1}\mathbf{F} + \mathbf{W}_T = (\mathbf{X}_{T-2}\mathbf{F} + \mathbf{W}_{T-1})\mathbf{F} + \mathbf{W}_T \\ &= \mathbf{X}_1\mathbf{F}^{T-1} + \sum_{t=2}^{T}\mathbf{W}_t\mathbf{F}^{T-t} \\ &= \sum_{t=T-D+1}^{T}\mathbf{W}_t\mathbf{F}^{T-t}. \end{aligned}$$

Since the model parameters $\hat{\lambda}_T$ are updated using $\sum_{t\le T}\vec{\varphi}_t = \mathbf{X}_T\vec{e}$, which are related to recent $\mathbf{W}_t$ for $t \ge T - D + 1$, the observations before $T - D + 1$ have smaller contributions to the update of the model parameters. Therefore, the algorithm is suited for online update of the model parameters that vary slower than D. In other words, if the process is time-homogeneous, the forward-only algorithm based online update procedure will approach the ML estimation of the model parameters.

In the beginning of the online update, the length of the observation sequence is short. The model parameters must be updated more frequently so that they converge quickly. Therefore, we assume that $l_1 \le l_2 \le \ldots \le l_N \le D$ are updating periods. After the quick updates, the model parameters will be updated in the interval D.

Algorithm 3.3 Forward-Only Based Online Updating Algorithm

1. *Set initial values:*
   set $\alpha_t(i,h)=0$ *for* $t<h,\ h\le D$;
   set $\mathrm{I}(o_{t-d+1:t}=v_{v_{k_1:k_d}})=0$ if $t-d+1<1$;
   let $T=1$; $\mathbf{X}_1^A=0$; $\mathbf{X}_1^B=\mathbf{I}_1 Diag(\vec{\alpha}_1)$;
2. *Online update with increasing intervals up to D:*
   for $n\ge 1$ {
   if $n>N$ { *let* $l_n=D$; }
   $t_1=T+1$ *and* $T=T+l_n$;
   2.1 Forward-only cumulating the smoothed probabilities:
   for $t=t_1,\ldots,T$ {
   if $o_{t+1:t+d}=v_{k_1:k_d}$ *is firstly observed* {
   let $b_{j,\,d}(v_{k_1:k_d})=0.5\times\min_{v_{k'_1:k'_d}}\{b_{j,d}(v_{k'_1:k'_d})>0\}$;
   }
   for $d=1,\ldots,D$ {
   compute $\mathbf{AB}'_{t+d}$;
   }
   let $\mathbf{Y}_t^A=\sum_{d=1}^{\min\{D,t-1\}}\mathbf{Z}_{t-d}\mathbf{B}'_t$ and $\mathbf{Y}_t^B=\mathbf{I}_t Diag(\vec{\alpha}_t)$;
   let $\mathbf{W}_t^A=(\mathbf{AB}'_{t+1}+\mathbf{Y}_t^A,\mathbf{AB}'_{t+2},\ldots,\mathbf{AB}'_{t+D})$ *and*
   $\mathbf{W}_t^B=(\mathbf{AB}'_{t+1}+\mathbf{Y}_t^B,\mathbf{AB}'_{t+2},\ldots,\mathbf{AB}'_{t+D})$;
   $\mathbf{X}_t^A=\mathbf{X}_{t-1}^A\mathbf{F}+\mathbf{W}_t^A$ *and* $\mathbf{X}_t^B=\mathbf{X}_{t-1}^B\mathbf{F}+\mathbf{W}_t^B$;
   }
   2.2 Obtain the expectation vectors:
   $\vec{g}_T^{\,A}=\mathbf{X}_T^A\vec{e}$ and $\vec{g}_T^{\,B}=\mathbf{X}_T^B\vec{e}$;
   2.3 Update the model parameters:
   for $i,j=1,\ldots,M$ *and* $h,d=1,\ldots,D$ {
   $k=(i-1)MD^2+(h-1)MD+(j-1)D+d$;
   $\hat{a}_{(i,h)(j,d)}=\vec{g}_T^{\,A}(k)$;
   }
   for $j=1,\ldots,M,\ d=1,\ldots,D,$ *and* $l=1,\ldots,L$ {
   $k=MD(l-1)+(j-1)D+d$ *and* $\hat{b}_{j,d}(v_{d,l})=\vec{g}_T^{\,B}(k)$;
   }
   }

3.4.2 Online Update by Maximizing Likelihood Function

As the length of the observation sequence increases, likelihood of the observation sequence is changing. Therefore, the purpose of the online update algorithm is to calculate the new likelihood and maximize it by updating the model parameters. Different from the usual online update algorithms found in the literature, this online update algorithm does not assume that the state ends at time T for an observation sequence $o_{1:T}$ and so the likelihood function is accurate.

Let $L_T(\lambda) = P[o_{1:T}|\lambda]$ be the likelihood function and λ be the set of model parameters to be found that maximizes $L_T(\lambda)$. Starting from an initial point of λ, the optimal point λ_T can be found that maximizes the likelihood function, that is, $\lambda_T = \arg\max_\lambda L_T(\lambda)$. We assume that the initial point is λ_{T-1}, which maximizes $L_{T-1}(\lambda)$. Therefore, λ_T is assumed to be close to λ_{T-1} for a time-homogeneous HSMM process.

The likelihood function for the observation sequence $o_{1:T}$ is

$$L_T(\lambda) = \sum_j \gamma_T(j) = \sum_j \sum_{t=T}^{T+D-1} \sum_{d=t-T+1}^{D} \alpha_t(j,d)$$

based on Eqn (2.13), where $\beta_t(j,d) = 1$, for $t \geq T$, according to the general assumption of boundary conditions of Eqn (2.9). It can be changed to

$$L_T(\lambda) = \sum_j \sum_{k=0}^{D-1} \sum_{d=k+1}^{D} \alpha_{T+k}(j,d) = \sum_j \sum_{k=0}^{D-1} \sum_{h=1}^{D-k} \alpha_{T+k}(j,h+k).$$

The forward formula (2.6) yields, for $k \geq 0$ and $D-k \geq d \geq 1$,

$$\alpha_{T+k}(j,d+k) = \sum_{i \neq j;h} \alpha_{T-d}(i,h) \cdot a_{(i,h)(j,d+k)} \cdot b_{j,d+k}(o_{T-d+1:T})$$

where $b_{j,d+k}(o_{T-d+1:T}) = \sum_{v_{d+1:d+k}} b_{j,d+k}(o_{T-d+1:T} v_{d+1:d+k})$ because the future observations $o_{T+1:T+k} = v_{d+1:d+k}$ could be any values. Denote the first part of the observable patterns $o_{T-d+1:T} v_{d+1:d+k}$ by $o_{[T-d+1:T}$. Then

$$L_T(\lambda) = \sum_j \sum_{k=0}^{D-1} \sum_{d=1}^{D-k} \sum_{i \neq j;h} \alpha_{T-d}(i,h) a_{(i,h)(j,d+k)} b_{j,d+k}(o_{[T-d+1:T}).$$

Because $\sum_{k=0}^{D-1}\sum_{d=1}^{D-k} = \sum_{d=1}^{D}\sum_{k=0}^{D-d} = \sum_{d=1}^{D}\sum_{\tau=d}^{D} = \sum_{\tau=1}^{D}\sum_{d=1}^{\tau}$ by letting $\tau = k + d$, we have

$$\begin{aligned} L_T(\lambda) &= \sum_{j,\tau}\sum_{d=1}^{\tau}\sum_{i \neq j;h} \alpha_{T-d}(i,h) a_{(i,h)(j,\tau)} b_{j,\tau}(o_{[T-d+1:T}) \\ &= \sum_{i;h}\sum_{j \neq i,\tau}\sum_{d=1}^{\tau}\sum_{v_{1:\tau}} \alpha_{T-d}(i,h) a_{(i,h)(j,\tau)} b_{j,\tau}(v_{1:\tau}) I(o_{[T-d+1:T} = v_{[1:d}), \end{aligned} \tag{3.16}$$

where $v_{1:\tau}$ are observable patterns of length τ, $v_{[1:d}$ their first part of length d, and I(x) is an indicator function which is 1 if x is true; otherwise 0.

To calculate $L_T(\lambda)$, we have to compute the forward variables $\alpha_{T-d}(i,h)$, for $D \geq h \geq 1$ and $D \geq d \geq 1$. When $T \leq D$, the observations $o_{1:T}$ are all taken into the computation. In this case, λ_T, for $T = 1,\ldots,D$, are the optimal solutions of $L_T(\lambda)$. When $T > D$, to compute the forward variables $\alpha_{T-d}(i,h)$, for $D \geq h \geq 1$ and $D \geq d \geq 1$, we have to compute the previous forward variables $\alpha_t(j,d)$ starting from $t = 1$. This is not suited for the online update of the model parameters when T becomes large. Therefore, some assumption must be made. Because o_T is the newest observation which brings a minor change in the model parameters for a time-homogenous HSMM, the previous forward variables $\alpha_t(j,d)$ for $t \leq T-1$ can be approximately assumed to be constant in finding λ_T. For example, the model parameters $a_{(i,h)(j,\tau)} \in \lambda$ can be updated by maximization of $L_T(\lambda)$. Their partial derivatives are simply

$$\frac{\partial L_T(\lambda)}{\partial a_{(i,h)(j,\tau)}} = \sum_{d=1}^{\tau}\sum_{v_{1:\tau}} \alpha_{T-d}(i,h) b_{j,\tau}(v_{1:\tau}) I(o_{[T-d+1:T} = v_{[1:d}).$$

The constraints are the normalization conditions $\sum_{j \neq i,\tau} a_{(i,h)(j,\tau)} = 1$ and $\sum_{v_{1:h}} b_{i,h}(v_{1:h}) = 1$, for all i and h. Suppose $b_{j,\tau}(v^*_{1:\tau})$ is a model parameter related to the newest observation o_T, that is, $\sum_{d=1}^{\tau} I(o_{[T-d+1:T} = v^*_{[1:d}) \neq 0$. Then $b_{j,\tau}(v^*_{1:\tau})$ in Eqn (3.16) can be similarly updated by maximization of $L_T(\lambda)$. Other observation probabilities $b_{j,\tau}(v_{1:\tau})$ that are not related to the newly obtained observation o_T, that is, $\sum_{d=1}^{T} I(o_{[T-d+1:T} = v_{[1:d)}) = 0$ cannot be updated at this time.

Let $\hat{\lambda}^{(0)} = \lambda_{T-1}$ and $l = 1$. Compute the incremental score vector by

$$\vec{\Delta}_{l-1} = \left(\frac{\partial L_T(\hat{\lambda}^{(l-1)})}{\partial \theta_1}, \frac{\partial L_T(\hat{\lambda}^{(l-1)})}{\partial \theta_2}, \ldots, \frac{\partial L_T(\hat{\lambda}^{(l-1)})}{\partial \theta_{|\lambda|}} \right).$$

Then the vector $\hat{\lambda}^{(l)}$ of the model parameters can be updated with a small step ε by letting

$$\hat{\lambda}^{(l)} = \hat{\lambda}^{(l-1)} + \varepsilon \vec{\Delta}_{l-1}$$

and $l = l + 1$. Repeat this iteration until the likelihood function $L_T(\lambda)$ converges/increases to a fixed point $L_T(\hat{\lambda}^{(l^*)})$, or the number of iterations reaches the limit $l_{\max^*}$. Let $\lambda_T = \hat{\lambda}^{(l^*)}$. Note that the limit of iteration number could be $l_{\max^*} = 1$ so that the computation amount at each time T is not too large.

Obviously, the new model parameters λ_T are determined mainly by the recent statistic characteristics of the stochastic process. The online update procedure is thus applicable to the situation that the model parameters are changing slowly in about D period.

3.4.3 Online Update for ML Segmentation

An observation sequence is assumed to be produced by a series of hidden states, each of which produces a sub-sequence or segment of observations. To divide the observation sequence into multiple segments/sub-sequences, we have to find the ML state sequence. Hence, the maximum log-likelihood of the observation sequence $o_{1:T}$ for segmentation is defined as (Squire and Levinson, 2005)

$$L_T(\lambda) \equiv \max_n \max_{\substack{d_1, \ldots, d_n \\ d_1 + \ldots d_n = T}} \log P[o_{1:T}, d_{1:n} | \lambda] = \max_n L_T^{(n)}(\lambda)$$

and

$$L_T^{(n)}(\lambda) \equiv \max_{\substack{d_1, \ldots, d_n \\ d_1 + \ldots + d_n = T}} \log P[o_{1:T}, d_{1:n} | \lambda],$$

where $d_{1:n} = (d_1, \ldots, d_n)$ denotes n segments, $d_k \in \mathbf{D}$ is the length of the kth segment, and $\sum_{k=1}^{n} d_k = T$. The corresponding state sequence is $(i_1, \ldots, i_n)$, the corresponding start times of the states are $\varsigma_1, \ldots, \varsigma_n$ with $\varsigma_1 = 1$, and the ending times are $\tau_1, \ldots, \tau_n$ with $\tau_n = T$. The probability $P[o_{1:T}, d_{1:n}|\lambda]$ can be decomposed into

$$
\begin{aligned}
P[o_{1:T}, d_{1:n}|\lambda] &= \sum_{i_n} P[o_{1:T}, d_{1:n-1}, (i_n, d_n)|\lambda] \\
&= \sum_{i_n} P[o_{1:\tau_{n-1}}, d_{1:n-1}|\lambda] P[(i_n, d_n), o_{\varsigma_n:\tau_n}|o_{1:\tau_{n-1}}, d_{1:n-1}, \lambda] \\
&= P[o_{1:\tau_{n-1}}, d_{1:n-1}|\lambda] \sum_{i_n} \alpha'_T(i_n, d_n),
\end{aligned}
$$

where the last state is assumed to end at $t = T$, and

$$
\begin{aligned}
\alpha'_T(i_n, d_n) &\equiv P[(i_n, d_n), o_{\varsigma_n:\tau_n}|o_{1:\tau_{n-1}}, d_{1:n-1}, \lambda] \\
&= \frac{P[(i_n, d_n), o_{\varsigma_{n-1}:\tau_n}, d_{n-1}|o_{1:\tau_{n-2}}, d_{1:n-2}, \lambda]}{P[o_{\varsigma_{n-1}:\tau_{n-1}}, d_{n-1}|o_{1:\tau_{n-2}}, d_{1:n-2}, \lambda]} \\
&= \frac{\sum_{i_{n-1}} P[(i_{n-1}, d_{n-1}), (i_n, d_n), o_{\varsigma_{n-1}:\tau_n}|o_{1:\tau_{n-2}}, d_{1:n-2}, \lambda]}{\sum_{i_{n-1}} P[(i_{n-1}, d_{n-1}), o_{\varsigma_{n-1}:\tau_{n-1}}|o_{1:\tau_{n-2}}, d_{1:n-2}, \lambda]} \\
&= \frac{\sum_{i_{n-1}} \alpha'_{\tau_{n-1}}(i_{n-1}, d_{n-1}) a_{(i_{n-1}, d_{n-1})(i_n, d_n)} b_{i_n, d_n}(o_{\varsigma_n:\tau_n})}{\sum_{i_{n-1}} \alpha'_{\tau_{n-1}}(i_{n-1}, d_{n-1})}.
\end{aligned}
$$

Denote $(i_{n-1}, d_{n-1}) = (i, h)$ and $(i_n, d_n) = (j, d)$, and assume that the optimal duration of the previous state has been determined, that is, $h = d^*_{T-d}$ and thus

$$
\alpha'_T(j, d) = \frac{\sum_i \alpha'_{T-d}(i, d^*_{T-d}) a_{(i, d^*_{T-d})(j, d)} b_{j,d}(o_{T-d+1:T})}{\sum_i \alpha'_{T-d}(i, d^*_{T-d})}.
$$

Then

$$L_T^{(n)}(\lambda) = \max_{\substack{d_1,\ldots,d_n \\ d_1+\cdots+d_n=T}} \left(\log P[o_{1:T-d_n}, d_{1:n-1}|\lambda] + \log \sum_{i_n} \alpha'_T(i_n, d_n) \right)$$

$$= \max_d \left(L_{T-d}^{(n-1)}(\lambda) + \log \sum_j \alpha'_T(j,d) \right).$$

Therefore, the log-likelihood becomes

$$\begin{aligned} L_T(\lambda) &= \max_n \max_d \left(L_{T-d}^{(n-1)}(\lambda) + \log \sum_j \alpha'_T(j,d) \right) \\ &= \max_d \max_n \left(L_{T-d}^{(n-1)}(\lambda) + \log \sum_j \alpha'_T(j,d) \right) \\ &= \max_d \left(L_{T-d}(\lambda) + \log \sum_j \alpha'_T(j,d) \right). \end{aligned} \tag{3.17}$$

Let

$$d_T^* = \arg\max_d \left(L_{T-d}(\lambda) + \log \sum_j \alpha'_T(j,d) \right)$$

be the optimal duration of the last state. Denote $d^* = d_T^*$ as the optimal duration of the last state and $h^* = d_{T-d^*}^*$ as the optimal duration of the state before the last state. Then

$$\alpha'_T(j,d^*) = \frac{\sum_i \alpha'_{T-d^*}(i,h^*) a_{(i,h^*)(j,d^*)} b_{j,d^*}(o_{T-d^*+1:T})}{\sum_i \alpha'_{T-d^*}(i,h^*)}. \tag{3.18}$$

Because only the last segment $(i_n, d_n) = (j, d^*)$ contains the new observation o_T, $\alpha'_T(j,d^*)$ is assumed being governed by the new model parameters and $L_{T-d^*}(\lambda)$ is assumed to be known. Therefore, to maximize the likelihood function $L_T(\lambda)$ with respect to λ, derivative of

$\alpha'_T(j, d^*)$ is required. The set of model parameters $\hat{\lambda}$ can be updated. That is, for each parameter θ_k in λ, at each time T, let

$$v_T^{(l)}(\lambda) \equiv \frac{\partial}{\partial \theta_k} \log \sum_j \alpha'_T(j, d^*)$$
$$= \frac{1}{\sum_j \alpha'_T(j, d^*)} \sum_j \frac{\partial}{\partial \theta_k} \alpha'_T(j, d^*),$$

where

$$\frac{\partial}{\partial \theta_k} \alpha'_T(j, d^*) = \frac{\sum_i \alpha'_{T-d^*}(i, h^*) \frac{\partial}{\partial \theta_k} \left[a_{(i,h^*)(j,d^*)} b_{j,d^*}(o_{T-d^*+1:T}) \right]}{\sum_i \alpha'_{T-d^*}(i, h^*)}.$$

After the incremental score vector $v_T(\lambda) = (v_T^{(1)}(\lambda), \ldots, v_T^{(|\lambda|)}(\lambda))$ is obtained, the model parameters λ can be updated with a small step $\varepsilon > 0$, that is, $\hat{\lambda} = \lambda + \varepsilon v_T(\lambda)$. Repeat this procedure at time T until the likelihood function converges to a fixed point.

Note that, in the update procedure at each time T, only a subset of model parameters, $a_{(i,h^*)(j,d^*)}$ for given h^* and d^*, and $b_{j,d^*}(o_{T-d^*+1:T})$ for given observable strings $v_{k_1:k_{d^*}} = o_{T-d^*+1:T}$ of length d^*, are updated. The updated probabilities and the un-updated ones must be normalized at each update step. For instance, suppose $\hat{a}_{(i,h^*)(j,d^*)}$ are the updated ones and $a_{(i,h)(j,d)}$ the un-updated ones. If $c = \hat{a}_{(i,h^*)(j,d^*)} + \sum_{(j,d) \neq (j,d^*)} a_{(i,h^*)(j,d)}$, then let $\hat{a}_{(i,h^*)(j,d^*)} = \hat{a}_{(i,h^*)(j,d^*)}/c$ and $a_{(i,h^*)(j,d)} = a_{(i,h^*)(j,d)}/c$. After normalized, the likelihood must be recomputed to make sure that the likelihood is increased.

[illegible] is required. The set of model parameters [illegible] can be updated [illegible] for each parameter [illegible]

$$[illegible]$$

where

$$[illegible]$$

[illegible] the incremental [illegible] obtained [illegible] the parameter [illegible] can be updated [illegible] [illegible] the likelihood [illegible]

[illegible]

CHAPTER 4

Implementation of HSMM Algorithms

This chapter discusses various practical problems and solutions in implementation of the general HSMM.

4.1 HEURISTIC SCALING

A general heuristic method to solve the underflow problem is to rescale the forward–backward variables by multiplying a large factor whenever an underflow is likely to occur (Levinson et al., 1983; Cohen et al., 1997).

Rewrite the forward recursion formula (2.6) as follows:

$$\tilde{\alpha}_t(j,d) = \sum_{i \in \mathbf{S} \setminus \{j\}} \sum_{h \in \mathbf{D}} \tilde{\alpha}_{t-d}(i,h) \cdot a_{(i,h)(j,d)} \cdot \tilde{b}_{j,d}(o_{t-d+1:t})$$

where $\tilde{\alpha}_t(j,d)$, for $j \in \mathbf{S}$ and $d \in \mathbf{D}$, are the scaled forward variables,

$$\tilde{b}_{j,d}(o_{t-d+1:t}) \equiv \left(\prod_{\tau=t-d+1}^{t} c_\tau \right) b_{j,d}(o_{t-d+1:t})$$

are the scaled observation probabilities, and c_t is the scale factor at time t. Note that for any state j, the scale factors are the same. Each scaled observation probability $\tilde{b}_{j,d}(o_{t-d+1:t})$ was multiplied by d scale factors corresponding to the observations $o_{t-d+1:t}$, for $d \in \mathbf{D}$. Based on the rewritten forward recursion formula, it can be proved that $\tilde{\alpha}_t(j,d) = \alpha_t(j,d) \prod_{\tau=1}^{t} c_\tau$. At each time t, the scale factor c_t can assume any value greater than zero, but usually takes the value such that $\sum_{j \in \mathbf{S}} \sum_{d \in \mathbf{D}} \tilde{\alpha}_t(j,d) = 1$, that is,

$$\frac{1}{c_t} = \sum_{j \in \mathbf{S}} \sum_{d \in \mathbf{D}} \left(\sum_{i \in \mathbf{S} \setminus \{j\}} \sum_{h \in \mathbf{D}} \tilde{\alpha}_{t-d}(i,h) a_{(i,h)(j,d)} \right) \left(\prod_{\tau=t-d+1}^{t-1} c_\tau \right) b_{j,d}(o_{t-d+1:t}). \tag{4.1}$$

Therefore, at each time t, $\tilde{\alpha}_t(j,d)$ is bounded by $\sum_{j\in\mathbf{S}}\sum_{d\in\mathbf{D}}\tilde{\alpha}_t(j,d)=1$. Similarly, the backward recursion formula (2.7) can be rewritten as

$$\tilde{\beta}_t(j,d)=\sum_{i\in\mathbf{S}\setminus\{j\}}\sum_{h\in\mathbf{D}}a_{(j,d)(i,h)}\tilde{b}_{i,h}(o_{t+1:t+h})\tilde{\beta}_{t+h}(i,h),$$

where $\tilde{\beta}_t(j,d)$, for $j\in\mathbf{S}$ and $d\in\mathbf{D}$, are the scaled backward variables. It can be derived that $\tilde{\beta}_t(j,d)=\beta_t(j,d)\prod_{\tau=t+1}^{T}c_\tau$. Note that the scale factors $\{c_t\}$ are determined in the forward recursion but not dedicated to the backward variables, and so cannot guarantee $\tilde{\beta}_t(j,d)$ to be finite or away from underflow problem, as discussed by Murphy (2002a).

By substituting the scaled forward and backward variables into Eqns (2.10)–(2.14), the scaled joint probabilities $\tilde{\eta}_t(j,d)$, $\tilde{\xi}_t(i,d';j,d)$, $\tilde{\xi}_t(i,j)$, and $\tilde{\gamma}_t(j)$ can be calculated. Denote $c=\prod_{t=1}^{T}c_t$ as the complex factor. Then it can be proved that $\tilde{\eta}_t(j,d)=c\eta_t(j,d)$, $\tilde{\xi}_t(i,h;j,d)=c\xi_t(i,h;j,d)$, $\tilde{\xi}_t(i,j)=c\xi_t(i,j)$, and $\tilde{\gamma}_t(j)=c\gamma_t(j)$. Because the complex factor c will appear for any i, j, h, and d, it will be eliminated when the model parameters are normalized, and so in the model parameter estimation formulas (3.4)–(3.6), those scaled joint probabilities can be directly substituted into these equations for the model parameter estimation.

In summary, except the observation probabilities $\tilde{b}_{j,d}(o_{t-d+1:t})$ are explicitly scaled and the scale factors are explicitly computed using Eqn (4.1), all the formulas originally used for the forward–backward recursions, the joint/posterior probability calculations, and the model parameter estimations can be directly applied without any change.

Because

$$\sum_{j\in\mathbf{S}}\tilde{\gamma}_t(j)=\sum_{j\in\mathbf{S}}\tilde{\gamma}_T(j)=\sum_{j\in\mathbf{S}}\sum_{\tau\geq T}\sum_{d=\tau-T+1}^{D}\tilde{\alpha}_\tau(j,d),$$

as given by Eqns (2.13) and (2.10), and

$$1=\sum_{\substack{j\in\mathbf{S}\\ d\in\mathbf{D}}}\tilde{\alpha}_T(j,d)\leq\sum_{j\in\mathbf{S}}\sum_{\substack{\tau\geq T\\ d\geq\tau-T+1}}\tilde{\alpha}_\tau(j,d)\leq\sum_{\substack{\tau\leq T+D-1\\ \tau\geq T}}\sum_{\substack{j\in\mathbf{S}\\ d\in\mathbf{D}}}\tilde{\alpha}_\tau(j,d)=D,$$

$\tilde{\gamma}_t(j)$ is bounded. Specially, under the simplified assumption of the boundary conditions that the last state ends at T, we have $\sum_{j\in\mathbf{S}}\tilde{\gamma}_t(j)=\sum_{j\in\mathbf{S}}\tilde{\gamma}_T(j)=\sum_{j\in\mathbf{S}}\sum_{d=1}^{D}\tilde{\alpha}_T(j,d)=1$, and $P[o_{1:T}|\lambda]=c^{-1}$.

In a multigram model, some of string patterns are observable, and others are not. If $o_{t-d+1:t}$ is an observable pattern of d-length string for given state j, then $b_{j,d}(o_{t-d+1:t})>0$; otherwise $b_{j,d}(o_{t-d+1:t})=0$. For example, in a text, words are observable string patterns. Suppose a sequence of observable patterns is $p_{1:n}=o_{1:T}$, where $o_{t-d:t-1}=p_l$ and $o_{t:t+k-1}=p_{l+1}$ are two observable patterns. Since $o_{t-d+1:t}$, for $d=1,\ldots, D$, are not observable patterns, the observation probabilities $b_{j,d}(o_{t-d+1:t})=0$, for all j and d. This results in that the scale factor $c_t=\infty$ based on Eqn (4.1). In this case, we can let the scale factor $c_t=1$. Then the forward–backward algorithm can continue without any change.

In fact, $\sum_{j\in \mathbf{S}}\sum_{d\in \mathbf{D}}\tilde{\alpha}_t(j,d)=1$ means any state must end at time t. This assumption is usually made in the literature. However, this may result in that the scale factors c_t selected by Eqn (4.1) is not appropriate in some special cases, as shown in Example 4.1.

Example 4.1 Failed Scaling

One might think that under the assumption of conditional independence of observations, there may be no scaling problem like the multigram model. But this is not always true, as shown in this example. Assume the observations are conditionally independent for given state, that is, $b_{j,d}(o_{t-d+1:t})=\prod_{\tau=t-d+1}^{t} b_j(o_\tau)$, for any j.

Suppose o_t is an observation at time t with $b_{j^*}(o_t)>0$ and $b_j(o_t)=0$, for all $j\neq j^*$, and o_{t-d^*} is an observation at time $t-d^*$ with $b_{j^*}(o_{t-d^*})=0$ and $b_j(o_{t-d^*})>0$, for all $j\neq j^*$, where j^* and d^* are given. Then $b_{j^*,d}(o_{t-d+1:t})=0$, for all $d>d^*$, and $b_{j,d}(o_{t-d+1:t})=0$, for all $j\neq j^*$ and all d. If state j^* takes long durations with high probabilities, that is, for any (i,h), there exist $a_{(i,h)(j^*,d)}\approx 0$, for all $d\le d^*$, then $a_{(i,h)(j^*,d)}b_{j^*,d}(o_{t-d+1:t})\approx 0$, for all d. In other words, state j^* cannot end at time t. Since $b_{j,d}(o_{t-d+1:t})=0$, for all $j\neq j^*$ and all d, other states cannot end at time t too, with $a_{(i,h)(j,d)}b_{j,d}(o_{t-d+1:t})=0$, for all $(i,h), j\neq j^*$ and d.

In this case, we can see that the scale factor given by Eqn (4.1) becomes $c_t^{-1}\approx 0$ or $c_t\approx\infty$. Due to

$$\begin{aligned}\tilde{\beta}_{t-1}(j,d)&=\sum_{h>d^*} a_{(j,d)(j^*,h)}\tilde{b}_{j^*,h}(o_{t:t+h-1})\tilde{\beta}_{t+h-1}(j^*,h)\\&=\sum_{h>d^*} a_{(j,d)(j^*,h)}\tilde{b}_{j^*,h-1}(o_{t+1:t+h-1})c_t b_{j^*}(o_t)\tilde{\beta}_{t+h-1}(j^*,h)\to\infty,\end{aligned}$$

the scaled backward variables $\tilde{\beta}_{t-1}(j,d)$ may become infinite. In other words, the scale factors cannot guarantee that the scaled backward variables are bounded.

Since $o_{t+1:t+k}$ are the future observations that can be any values and $b_{j^*,d+k}(o_{t-d+1:t+k}) = b_{j^*,d}(o_{t-d+1:t}) > 0$, for all $d \le d^*$ and $k \ge 0$, $a_{(i,h)(j^*,d+k)} b_{j^*,d+k}(o_{t-d+1:t+k}) > 0$, for all $k + d > d^*$. Therefore, if c_t is selected such that $\sum_{j \in \mathbf{S}} \tilde{\gamma}_t(j) = 1$ instead of $\sum_{j \in \mathbf{S}} \sum_{d \in \mathbf{D}} \tilde{\alpha}_t(j,d) = 1$ by assuming that the state is not necessarily ending at t (Li and Yu, 2015), then

$$c_t^{-1} \equiv \sum_{j \in \mathbf{S}} \sum_{\substack{d,k \\ d+k \in \mathbf{D}}} \left(\sum_{i \in \mathbf{S} \setminus \{j\}} \sum_{h \in \mathbf{D}} \tilde{\alpha}_{t-d}(i,h) a_{(i,h)(j,d+k)} \right) \left(\prod_{\tau=t-d+1}^{t-1} c_\tau \right) b_{j,d}(o_{t-d+1:t})$$

will not be zero.

4.2 POSTERIOR NOTATION

This section uses the notation of posterior probabilities to overcome the underflow problem involved in the recursive calculation of the joint probabilities associated with a sequence of observations (Guedon and Cocozza-Thivent, 1989; Yu and Kobayashi, 2006). Because the posterior probabilities always satisfy the normalization conditions and will not decay with the increase of the observation sequence length, the refined forward–backward algorithm for the HSMM becomes robust against the underflow problem, without increasing the computational complexity. This is similar to the standard HMM whose forward–backward algorithms can automatically avoid the underflow problem by replacing the joint probabilities with conditional probabilities (Devijver, 1985; Askar and Derin, 1981). However, in the HSMM, the likelihood function $P[o_{1:t}|\lambda]$ for the partial observations $o_{1:t}$ is difficult to be calculated, which will result in that the forward–backward algorithm for the HSMM must be changed a lot.

To be conditioned on observations, the forward variables are redefined in the notation of posterior probabilities by

$$\overline{\alpha}_t(j,d) \equiv P[S_{[t-d+1:t]} = j | o_{1:t}, \lambda] = \frac{\alpha_t(j,d)}{P[o_{1:t}|\lambda]},$$

with the initial values $\overline{\alpha}_t(j,d) = \pi_{j,d}$ for $t \le 0$, $d \in \mathbf{D}$ and $j \in \mathbf{S}$. To keep the form of the smoothed probabilities $\frac{\eta_t(j,d)}{P[o_{1:T}|\lambda]} = \frac{\alpha_t(j,d)\beta_t(j,d)}{P[o_{1:T}|\lambda]} = \overline{\alpha}_t(j,d)\overline{\beta}_t(j,d)$, the backward variables are redefined as the ratio to the corresponding predicted probabilities by

$$\overline{\beta}_t(j,d) = \frac{P[o_{1:t}|\lambda]\beta_t(j,d)}{P[o_{1:T}|\lambda]} = \frac{\beta_t(j,d)}{P[o_{t+1:T}|o_{1:t},\lambda]} = \frac{P[o_{t+1:T}|S_{[t-d+1:t]}=j,\lambda]}{P[o_{t+1:T}|o_{1:t},\lambda]},$$

with initial values $\overline{\beta}_T(j,d)=1$, for all $d\in\mathbf{D}$ and $j\in\mathbf{S}$. The posterior forms of observation probabilities are defined by

$$\overline{b}_{j,d}(o_{t-d+1:t}) \equiv \frac{b_{j,d}(o_{t-d+1:t})}{P[o_{t-d+1:t}|o_{1:t-d},\lambda]}. \tag{4.2}$$

Based on these new definitions, $\overline{\alpha}_t(j,d)$, $\overline{\beta}_t(j,d)$, and $\overline{b}_{j,d}(o_{t-d+1:t})$ will not decay with increase of time t. By substituting these new definitions into the forward–backward formulas (2.6) and (2.7), the posterior form of the forward–backward formulas are yielded as follows:

$$\overline{\alpha}_t(j,d) = \sum_{i\neq j;h} \overline{\alpha}_{t-d}(i,h)a_{(i,h)(j,d)}\overline{b}_{j,d}(o_{t-d+1:t}) \tag{4.3}$$

and

$$\overline{\beta}_t(j,d) = \sum_{i\neq j;h} a_{(j,d)(i,h)}\cdot\overline{b}_{i,h}(o_{t+1:t+h})\cdot\overline{\beta}_{t+h}(i,h). \tag{4.4}$$

They have the same forms as the original forward–backward formulas (2.6) and (2.7).

To compute the posterior forms of observation probabilities by Eqn (4.2), we have to compute $P[o_{1:t}|\lambda]/P[o_{1:t-d}|\lambda]$. The likelihood function $P[o_{1:t}|\lambda]$ for the partial observations $o_{1:t}$ can be iteratively determined by

$$\begin{aligned}
P[o_{1:t}|\lambda] &= \sum_{j\in\mathbf{S}} P[S_t=j,o_{1:t}|\lambda] \\
&= \sum_{j\in\mathbf{S}} \sum_{\substack{d\geq 1\\ k\geq 0\\ d+k\leq D}} P[S_{[t-d+1:t+k]}=j,o_{1:t}|\lambda] \\
&= \sum_{j\in\mathbf{S}} \sum_{\substack{d\geq 1\\ k\geq 0\\ d+k\leq D}} \sum_{i\in\mathbf{S}\setminus\{j\}} \sum_{h\in\mathbf{D}} \alpha_{t-d}(i,h)a_{(i,h)(j,d+k)}b_{j,d+k}(o_{t-d+1:t}) \\
&= \sum_{\substack{d\geq 1\\ k\geq 0\\ d+k\leq D}} P[o_{1:t-d}|\lambda] \sum_{j\in\mathbf{S}} \sum_{i\in\mathbf{S}\setminus\{j\}} \sum_{h\in\mathbf{D}} \overline{\alpha}_{t-d}(i,h)a_{(i,h)(j,d+k)}b_{j,d+k}(o_{t-d+1:t}),
\end{aligned} \tag{4.5}$$

with initial values $P[o_{1:t-d}|\lambda]=1$ for $t \leq d$, where $b_{j,d+k}(o_{t-d+1:t}) = \sum_{o_{t+1:t+k}} b_{j,d+k}(o_{t-d+1:t+k})$ because the future observations $o_{t+1:t+k}$ could be any values. Denote $C(t_1, t_2) = P[o_{1:t_1}|\lambda]/P[o_{1:t_2}|\lambda]$. If $t_1 > t_2$, $C(t_1, t_2) = P[o_{t_2+1:t_1}|o_{1:t_2}, \lambda]$; if $t_1 = t_2$, $C(t_1, t_2) = 1$; if $t_1 < t_2$, $C(t_1, t_2) = 1/P[o_{t_1+1:t_2}|o_{1:t_1}, \lambda]$. Then Eqn (4.5) becomes

$$\begin{aligned} C(t, t-\tau) &= \frac{P[o_{1:t}|\lambda]}{P[o_{1:t-\tau}|\lambda]} \\ &= \sum_{\substack{d \geq 1 \\ k \geq 0 \\ d+k \leq D}} C(t-d, t-\tau) \sum_{j \in \mathbf{S}} \sum_{i \in \mathbf{S} \setminus \{j\}} \sum_{h \in \mathbf{D}} \overline{\alpha}_{t-d}(i,h) a_{(i,h)(j,d+k)} b_{j,d+k}(o_{t-d+1:t}), \end{aligned}$$

for $\tau \in \mathbf{D}$, where the initial values are $C(t_1, t_2) = 1$ when both $t_1 \leq 0$ and $t_2 \leq 0$. In other words, we can iteratively compute the scale factors $C(t, t-d)$. Then the observation probabilities can be scaled by letting $\overline{b}_{j,d}(o_{t-d+1:t}) = b_{j,d}(o_{t-d+1:t})/C(t, t-d)$ at each time t.

Usually in the literature, it is assumed that the last state ends at time t, then the calculation of Eqn (4.5) becomes simpler. However, this assumption may not be appropriate in some special cases, as shown in Example 4.1.

After the forward–backward variables are determined, the smoothed probabilities determined by Eqns (2.10)–(2.14) can be yielded, such as

$$\overline{\eta}_t(j,d) \equiv P\left[S_{[t-d+1:t]} = j | o_{1:T}, \lambda\right] = \frac{\eta_t(j,d)}{P[o_{1:T}|\lambda]} = \overline{\alpha}_t(j,d)\overline{\beta}_t(j,d)$$

and

$$\begin{aligned} \overline{\xi}_t(i,h;j,d) &= P[S_{[t-h+1:t]} = i, S_{[t+1:t+d]} = j | o_{1:T}, \lambda] \\ &= \frac{\xi_t(i,h;j,d)}{P[o_{1:T}|\lambda]} \\ &= \overline{\alpha}_t(i,h) a_{(i,h)(j,d)} \overline{b}_{j,d}(o_{t+1:t+d}) \overline{\beta}_{t+d}(j,d), \end{aligned}$$

which have the same forms as Eqns (2.10)–(2.14).

Compared with the heuristic scaling, the posterior notation has a solid background of theory. In the heuristic scaling method, the purpose is to scale the forward variables, and hence the backward variables cannot be guaranteed to be properly scaled. In the posterior notation method, both the forward and backward variables are properly scaled, with explicit posterior notion.

4.3 LOGARITHMIC FORM

If the forward–backward algorithm is implemented in the logarithmic domain, like the MAP and Viterbi algorithms used for turbo-decoding in digital communications, then the multiplications involved in computing the joint probabilities of observations become additions. The product x that may cause a floating-point underflow (e.g., 10^{-1023}) becomes $\ln x = -1023$ and finite. From the following algorithm we can see that the logarithmic values do not need to be transformed back in the forward–backward procedure. Therefore, the scaling of the forward–backward variables becomes unnecessary. Denote

$$\begin{aligned}
\dot{\alpha}_t(j,d) &\equiv \ln \alpha_t(j,d), \\
\dot{\beta}_t(j,d) &\equiv \ln \beta_t(j,d), \\
\dot{a}_{(i,h)(j,d)} &\equiv \ln a_{(i,h)(j,d)}, \\
\dot{b}_{j,d}(o_{t-d+1:t}) &\equiv \ln b_{j,d}(o_{t-d+1:t}), \\
\dot{\delta}_t(j,d) &\equiv \ln \delta_t(j,d), \\
\dot{\varepsilon}_t(j,d) &\equiv \ln \varepsilon_t(j,d),
\end{aligned}$$

and, for simplicity of notation, let

$$\begin{aligned}
\dot{\zeta}_t(i,h;j,d) &\equiv \ln P[S_{[t-d-h+1:t-d]} = i, S_{[t-d+1:t]} = j, o_{1:t}|\lambda] \\
&= \dot{\alpha}_{t-d}(i,h) + \dot{a}_{(i,h)(j,d)} + \dot{b}_{j,d}(o_{t-d+1:t}),
\end{aligned}$$

$$\begin{aligned}
\dot{\varsigma}_t(j,d;i,h) &\equiv \ln P[S_{[t+1:t+h]} = i, o_{t+1:T}|S_{[t-d+1:t]} = j, \lambda] \\
&= \dot{a}_{(j,d)(i,h)} + \dot{b}_{i,h}(o_{t+1:t+h}) + \dot{\beta}_{t+h}(i,h),
\end{aligned}$$

$$\dot{\Delta}_t(j,d) \equiv \max_{i \in \mathbf{S} \setminus \{j\}, h \in \mathbf{D}} \{\dot{\zeta}_t(i,h;j,d)\},$$

and

$$\dot{E}_t(j,d) \equiv \max_{i \in \mathbf{S} \setminus \{j\}, h \in \mathbf{D}} \{\dot{\varsigma}_t(j,d;i,h)\}.$$

Then the forward formula (2.6) can be rewritten in the logarithmic form as

$$\begin{aligned}\dot{\alpha}_t(j,d) &= \ln \sum_{i\in\mathbf{S}\backslash\{j\}} \sum_{h\in\mathbf{D}} \exp[\dot{\zeta}_t(i,h;j,d)] \\ &= \dot{\Delta}_t(j,d) + \ln \sum_{i\in\mathbf{S}\backslash\{j\}} \sum_{h\in\mathbf{D}} \exp[\dot{\zeta}_t(i,h;j,d) - \dot{\Delta}_t(j,d)],\end{aligned} \tag{4.6}$$

for $1 \le t \le T$, $j\in\mathbf{S}, d\in\mathbf{D}$, and the backward formula (2.7) as

$$\begin{aligned}\dot{\beta}_t(j,d) &= \ln \sum_{i\in\mathbf{S}\backslash\{j\}} \sum_{h\in\mathbf{D}} \exp[\dot{\varsigma}_t(j,d;i,h)] \\ &= \dot{E}_t(j,d) + \ln \sum_{i\in\mathbf{S}\backslash\{j\}} \sum_{h\in\mathbf{D}} \exp[\dot{\varsigma}_t(j,d;i,h) - \dot{E}_t(j,d)],\end{aligned} \tag{4.7}$$

for $0 \le t \le T-1$. Following Boutillon et al. (2003), an ADD operation is performed by using the Jacobi logarithm:

$$\ln(e^x + e^y) = \max(x,y) + \ln(1 + e^{-|x-y|}),$$

which is called the $MAX^*(x,y)$ operation, that is, a maximum operator adjusted by a correction factor. The second term, a function of the single variable $|x-y|$, can be precalculated and stored in a small lookup table (LUT) (Boutillon et al., 2003). As an extension of the operation, we denote

$$\begin{aligned}MAX^*\{w,x,y,z,\ldots\} &= \ln(e^w + e^x + e^y + e^z + \cdots) \\ &= \ln(e^{\ln(e^w+e^x)} + e^{\ln(e^y+e^z)} + \cdots) \\ &= \ln(e^{MAX^*(w,x)} + e^{MAX^*(y,z)} + \cdots) \\ &= MAX^*\{MAX^*(w,x), MAX^*(y,z),\ldots\}.\end{aligned}$$

That is, if there are n variables, $(n-1)$ operations of $MAX^*(x,y)$ are required for the final result. The $MAX^*(x,y)$ operation is simple and all those operations need only one common LUT. Then

$$\dot{\alpha}_t(j,d) = MAX^*\{\dot{\zeta}_t(i,h;j,d) : i\in\mathbf{S}\backslash\{j\}, h\in\mathbf{D}\},$$

for $1 \le t \le T$, $j\in\mathbf{S}, d\in\mathbf{D}$, and

$$\dot{\beta}_t(j,d) = MAX^*\{\dot{\varsigma}_t(j,d;i,h) : i\in\mathbf{S}\backslash\{j\}, h\in\mathbf{D}\},$$

for $0 \le t \le T-1$.

Obviously, in Eqns (4.6) and (4.7), we have $0 \le \ln\sum_{i\in\mathbf{S}}\sum_{h\in\mathbf{D}}\exp[\dot{x}(i,h) - \dot{x}_{\max}] \le \ln(MD)$. If $|\dot{\Delta}_t(j,d)|$ and $|\dot{E}_t(j,d)|$ are much greater than this limited value, then we can omit it and get $\dot{\alpha}_t(j,d) \approx \dot{\Delta}_t(j,d)$ and $\dot{\beta}_t(j,d) \approx \dot{E}_t(j,d)$. In this case, the logarithmic form of the forward–backward algorithm given by Eqns (4.6) and (4.7) is reduced to the logarithmic form of the Viterbi HSMM algorithm if $\dot{\alpha}_t(j,d)$ and $\dot{\beta}_t(j,d)$ are replaced with $\dot{\delta}_t(j,d)$ and $\dot{\varepsilon}_t(j,d)$, respectively. That is, they are equivalent to the logarithmic form of the Viterbi HSMM algorithm given by Eqns (2.29) and (2.30):

$$\dot{\delta}_t(j,d) \equiv \max_{i\in\mathbf{S}\setminus\{j\},h\in\mathbf{D}} \{\dot{\delta}_{t-d}(i,h) + \dot{a}_{(i,h)(j,d)} + \dot{b}_{j,d}(o_{t-d+1:t})\}$$

and

$$\dot{\varepsilon}_t(j,d) \equiv \max_{i\in\mathbf{S}\setminus\{j\},h\in\mathbf{D}} \{\dot{a}_{(j,d)(i,h)} + \dot{b}_{i,h}(o_{t+1:t+h}) + \dot{\varepsilon}_{t+h}(i,h)\}.$$

From this point of view, the bidirectional Viterbi HSMM algorithm can be considered an approximation of the forward–backward HSMM algorithm. For the smoothed probabilities, $\ln\eta_t(j,d) = \dot{\alpha}_t(j,d) + \dot{\beta}_t(j,d)$ and $\ln\xi_t(i,h;j,d) = \dot{\alpha}_t(i,h) + \dot{\varsigma}_t(i,h;j,d)$ can be approximately expressed by the variables of the bidirectional Viterbi HSMM algorithm. For example, let $\ln\eta_t(j,d) \approx \dot{\delta}_t(j,d) + \dot{\varepsilon}_t(j,d)$. The model parameters can also be approximately estimated using the bidirectional Viterbi HSMM algorithm.

4.4 PRACTICAL ISSUES IN IMPLEMENTATION

Other than the underflow problem, there exist several other practical issues in implementation of an HSMM.

4.4.1 Nonindexable Observables

Observations, $o_1, o_2, \ldots, o_T$, collected in practice may not be scalar values or integers. $\mathbf{V} = \{o_1, o_2, \ldots, o_T\}$ can be a set of symbols, events, signals, vectors, discrete or continuous values. For example, heads and tails of coin toss outcomes are observable symbols, and arrivals and departures of a queue are observable events. If the set of observables $\mathbf{V}$ contains only a few of discrete values, it is easy to map each of the values into an index, so that the observation probabilities $b_j(o_t)$ can be tabulated in the implementation of HSMM algorithms. However, it is often that $\mathbf{V}$ contains too many or uncountable values. To make the observation probability distributions $b_{j,d}(v_{k_1:k_d})$ easier to be determined in the implementation of HSMM algorithms, preprocessing of the observables is often required.

One usually divides the observation space into equally large areas/intervals. However, this method may not be appropriate to the case that a lot of observable values are very small (e.g., about 0.01) while a few of them are very large (e.g., 1000). In this case, some of the intervals may contain too many observable values while others contain few/zero values. Therefore, a reasonable and general method for preprocessing the observations is to separate equal number/area of observable values of the observation space into each interval. In other words, in the scalar case, we uniformly divide the cumulative distribution of **V** into equal intervals. Suppose $W(v) \in (0, 1]$ is the cumulative distribution of **V** and the observable values are to be divided into K intervals. Then let the kth interval be $(x_{k-1}, x_k]$, where $x_k = \max\{v: W(v) \leq k/K, v \in \mathbf{V}\}$, $k = 0, \ldots, K$. Therefore, $o_t \in (x_{k-1}, x_k]$ is mapped into integer k.

The total number of intervals, K, should be selected such that every interval contains at least one observable value but not too many. Note that in the cumulative distribution of **V**, each observable is treated as having the equal probability $1/|\mathbf{V}|$, but not its real probabilities $b_j(v_k)$ that it is produced in given states. Therefore, the observations in different intervals may have different observation probabilities for a given state.

Example 4.2 Preprocessing Observations

Suppose the observation sequence is $o_{1:20}$, where

o_1	o_2	o_3	o_4	o_5	o_6	o_7	o_8	o_9	o_{10}
1.2	*32.7*	*1.5*	*912.6*	*1.3*	*1.4*	*1.4*	*1.5*	*9.4*	*262.1*
o_{11}	o_{12}	o_{13}	o_{14}	o_{15}	o_{16}	o_{17}	o_{18}	o_{19}	o_{20}
9.2	*9.2*	*1.2*	*9.5*	*262.1*	*9.5*	*1.3*	*1.2*	*1.3*	*9.5*

The set of observable values is thus $\mathbf{V} = \{1.2, 32.7, 1.5, 912.6, 1.3, 1.4, 9.4, 262.1, 9.2, 9.5\}$. Sort **V** and get $\mathbf{V} = \{1.2, 1.3, 1.4, 1.5, 9.2, 9.4, 9.5, 32.7, 262.1, 912.6\} = \{v_1, v_2, \ldots, v_{10}\}$. The cumulative distribution of **V** is $W(v) = (1, 2, \ldots, 10)/10$. If the observable values are to be mapped into $K = 8$ integers, then let $x_k = \max\{v: W(v) \leq k/8, v \in \mathbf{V}\}$, for $k = 0, \ldots, 8$. We get $x_0 = 0$, $x_1 = v_1$, $x_2 = v_2$, $x_3 = v_3$, $x_4 = v_5$, $x_5 = v_6$, $x_6 = v_7$, $x_7 = v_8$, $x_8 = v_{10}$. Now map $1.2 \to 1$, $1.3 \to 2$, $1.4 \to 3$, $(1.5, 9.2) \to 4$, $9.4 \to 5$, $9.5 \to 6$, $32.7 \to 7$, $(262.1, 912.6) \to 8$, as shown in Figure 4.1. Finally, the observations are mapped into $o'_{1:20} = (1, 7, 4, 8, 2, 3, 3, 4, 5, 8, 4, 4, 1, 6, 8, 6, 2, 1, 2, 6)$.

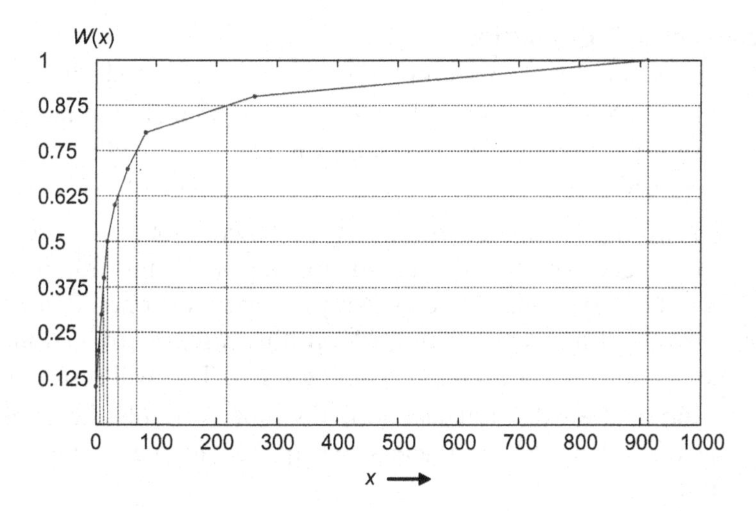

Figure 4.1 $W(x)$, the cumulative distribution of V.
The x-axis represents the observable values, and the y-axis is the normalized cumulative number or integral of the observable values. If the observable values are discrete, then W(x) represents the cumulative number of observables normalized by the total number of observables; if the observable is continuous, then W(x) represents the integral normalized by the total integral. Divide the y-axis with equal interval such that in each of the corresponding intervals in the x-axis there is almost the same number/area of observable values.

Example 4.3 MATLAB Codes for Preprocessing Observations

The MATLAB codes for preprocessing the observation sequence is quite simple, such as follows:

$[V, I] = sort(O)$;	% sort the observation
$V = diff([0; V])$;	% find the same observable values
$V_k = V(V > 0)$;	% get the set of observable values
If $length(V_k) < K$	% compare K with the number of observables
$K = length(V_k)$;	% let $K \leq \|V_k\|$
end	
$V(V > 0) = 1$;	% get the cumulative distribution of observables
$V = cumsum(V)$;	
$V = floor(V./\max(V).*K) + 1$;	% divide the observables into K periods
$O(I) = V$;	% Map the observations into integers

where O is a $T \times 1$ observation sequence, V_k is the set of observable values, and $O(I) = V$ is the observations mapped into $1{:}K$ indexes.

4.4.2 Missing Observables

Suppose a training set of samples is $o_1, o_2, \ldots, o_T$, and the set of observables is $\mathbf{V} = \bigcup_{t=1}^{T}\{o_t\}$. After the set of model parameters is trained using the training set, the trained model is applied to a testing set of observations, $o'_1, o'_2, \ldots, o'_{T'}$, with a set of observables $\mathbf{V}' = \bigcup_{t=1}^{T'}\{o'_t\}$. Because the testing set of observations may contain some rare observable values that did not appear in the training set, $\mathbf{V} \supseteq \mathbf{V}'$ is not always satisfied. This is a practical issue one often faces. In other words, it may exist that $o'_t \notin \mathbf{V}$. In this case, the trained model parameters $b_{j,d}(o'_{t_1:t_2})$, for all $t_1 \leq t \leq t_2$, are zeros. This will result in the forward and backward variables and many other variables being zeros when the trained model parameters are applied to the testing set of observations.

To solve this problem, it is better to update the model parameters. The trained model parameters are used as the initial parameters, and all the zero parameters $b_{j,d}(o'_{t_1:t_2})$, for $t_1 \leq t \leq t_2$, are assumed with values as small as possible. Then use both $o_1, o_2, \ldots, o_T$ and $o'_1, o'_2, \ldots, o'_{T'}$ to update the model parameters. The updated model parameters are eventually applied to the testing set of observations.

Sometimes, the training set may not be kept for the testing period, or the testing set is very small, so that the update of the model parameters is not possible. In this case, a reasonable solution is letting $b_{j,d}(o'_{t_1:t_2})$ equal a small number, for example, letting $b_{j,d}(o'_{t_1} \cdots o'_t \cdots o'_{t_2}) = r \cdot \sum_{O_t} b_{j,d}(o'_{t_1} \cdots O_t \cdots o'_{t_2})$, with $0 < r \ll 1$.

To solve this problem on a solid background of theory, the methods proposed in HDP-HMMs (hierarchical Dirichlet process hidden Markov models) can be applied, where the number of distinct observables is expected to increase as the logarithm of the number of observations, and the probability of an entirely new observable is proportional to a parameter γ^e that can be learned from the given sequence of observations (Beal et al., 2002). Suppose $o'_{t'}$ is a new observable that never occurred before and the set of observables obtained before time t' is $\mathbf{U} = \mathbf{V} \cup \bigcup_{t=1}^{t'-1}\{o'_t\} = \{v_1, \ldots, v_K\}$, where $\mathbf{V} = \bigcup_{t=1}^{T}\{o_t\}$ is the set of observables appearing in the training set, and $\mathbf{U} \cap o'_{t'} = null$. Then the observation probabilities for handling the new observable $o'_{t'}$ can be updated by

$$b_j(o'_{t'}) \leftarrow \frac{\gamma^e}{T + t' - 1 + \gamma^e},$$

$$b_j(v_k) \leftarrow \frac{T + t' - 1}{T + t' - 1 + \gamma^e} b_j(v_k), \quad v_k \in \mathbf{U},$$

$$\mathbf{U} \leftarrow \mathbf{U} \cup \{o'_{t'}\}.$$

For instance, $\gamma^e = 1$.

4.4.3 Unknown Model Order

Even though the model is known, it is often in practice that the model order, that is, the total number of states, M, and the maximum length of state duration, D, is unknown. To determine the order of the model, a Bayesian approach (Ghahramani, 2001) can be used to estimate the unknown quantities of M and D. However, one may demand a simple way to get the model order quickly.

A simple method to find out the order of an HSMM is to try various values of M and D. Denote $\lambda^{(M,D)}$ as the model parameter with order M and D. The likelihood function is $P[o_{1:T}|\lambda^{(M,D)}]$. Then the ML estimation of $\lambda^{(M,D)}$ is

$$\hat{\lambda}^{(M,D)} = \arg\max_{\lambda^{(M,D)}} \log P[o_{1:T}|\lambda^{(M,D)}],$$

which can be determined using the re-estimation algorithms for given M and D. Then the order estimators given by Ephraim and Merhav (2002) can be used in the selection of the model order. For example, the order estimator proposed by Finesso (1990) can be used as the objective function for the selection of the model order:

$$(\hat{M}, \hat{D}) = \min\left\{\arg \min_{M>1, D\geq 1} \left\{-\frac{1}{T} \log P\left[o_{1:T}|\hat{\lambda}^{(M'D)}\right] + 2c_{MD}^2 \frac{\log T}{T}\right\}\right\},$$

where $c_{MD} = MD(MD + K - 2)$ is a penalty term that favors simpler models over more complex models, T is the total number of observations, K is the total number of values that an observation can take,

M is the total number of states, D is the maximum duration of a state, and so c_{MD} is the total number of free model parameters. Compared with the popular Akaike information criterion (AIC) (Akaike, 1974) and the Bayesian information criterion (BIC) (Schwarz, 1978), which use c_{MD} as the penalty, this is a strongly consistent estimator with stronger penalty c_{MD}^2.

A simpler method is to try a series of M and D, for $(M_1, D_1) \leq \cdots \leq (M_N, D_N)$, and get $\lambda^{(M_1,D_1)}, \ldots, \lambda^{(M_N,D_N)}$ using the re-estimation algorithms. Then the model order can be simply selected by finding the maximum of their likelihoods, that is, $(\hat{M}, \hat{D}) = \arg\max_n\{P[o_{1:T}|\lambda^{(M_n,D_n)}]\}$. It can also be selected by finding the point that the likelihood will not be improved significantly, that is, $(\hat{M}, \hat{D}) = (M_n, D_n)$ if $P[o_{1:T}|\lambda^{(M_1,D_1)}] < \ldots < P[o_{1:T}|\lambda^{(M_n,D_n)}]$ and $P[o_{1:T}|\lambda^{(M_n,D_n)}] \geq P[o_{1:T}|\lambda^{(M_{n+1},D_{n+1})}]$.

The simplest way is to set M and D being a little bit large, and then delete the nonappeared states and durations. For instance, if the estimated state sequence is $(\hat{s}_1, \hat{s}_2, \ldots, \hat{s}_T)$, let the set of states be $\mathbf{S} = \bigcup_{t=1}^{T}\{\hat{s}_t\}$. The maximum duration D can be determined by $\max_d \hat{s}_{[t-d+1,t]} = j$.

A theoretic way is assuming the state space infinite and the states corresponding to the given observation sequence are a finite number of samples of the state space. The HDP-HSMM (Johnson and Willsky, 2013; Nagasaka et al., 2014) is such a model that extends a nonparametric HSMM into a HDP mixture model, and provides a method that does not require an explicit parametric prior on the number of states or use model selection methods to select a fixed number of states. According to the theory of HDP, the number of instantiated states is gradually increased by sampling from the set of Dirichlet process distributions. In the sampling procedure, a previously appeared state is selected with the probability proportional to its counts having been transited from other states, and a new state is selected with the probability proportional to a given parameter.

4.4.4 Unknown Observation Distribution

Usually, based on the empirical knowledge on the stochastic process, the observation distribution $b_{j,d}(v_{k_1:k_d})$ can be determined whether they are parametric or nonparametric. If they are assumed to be parametric, their probability density distribution functions can be correspondingly determined. When the parametric distribution is unknown, the most popular ones that are often used in practice are a mixture of Gaussian distributions.

Example 4.4 Parametric Distribution of Observations

Use Example 1.4. Assume the observation distributions are parametric, and the request arrivals is characterized as a Poisson process modulated by an underlying (hidden state) semi-Markov process. The finite number of discrete states are defined by the discrete mean arrival rates. Let μ_j be the mean arrival rate for given state $j \in \mathrm{S}$. Then the number of arrivals in a time interval and the Markov state are related through the conditional probability distribution

$$b_j(k) = \frac{\mu_j^k}{k!} e^{-\mu_j},$$

where $b_{j,d}(v_{k_1:k_d})$ is assumed conditionally independent.

Note that when the observation distributions are parametric, the new parameters $\hat{\boldsymbol{\theta}}_j$ for state j can be found by maximizing $f(\boldsymbol{\theta}_j) \equiv \sum_{v_{k_1},\ldots,v_{k_d}} \hat{b}_{j,d}(v_{k_1:k_d}) \log b_{j,d}(v_{k_1:k_d}; \boldsymbol{\theta}_j)$ subject to the constraint $\sum_{v_{k_1},\ldots,v_{k_d}} b_{j,d}(v_{k_1:k_d}; \boldsymbol{\theta}_j) = 1$. For instance, if the probability density function $b_j(v_k)$, for $v_k = 0, 1, \ldots, \infty$, is Poisson with mean μ_j, then the parameter μ_j can be estimated by $\hat{\mu}_j = \sum_k \hat{b}_j(k)k$ or, equivalently, $\hat{\mu}_j = \sum_{t=1}^{T} \gamma_t(j) o_t / \sum_j \sum_{t=1}^{T} \gamma_t(j)$.

4.4.5 Unknown Duration Distribution

If the duration distributions $p_j(d)$ are required and unknown, then it is firstly required to determine whether they are nonparametric or parametric, depending on the specific preference of the applications. Among the parametric distributions, the most popular ones are the exponential family distribution, such as Poisson, exponential, Gaussian, and a mixture of Gaussian distributions. Generally, the Coxian distribution of duration can represent any discrete probability density function, and the underlying series–parallel network of the Coxian distribution also reveals the structure of different HSMMs.

4.4.6 Unordered States

Usually, an initial model parameter set λ_0 can be assumed based on empirical knowledge about the model. Sometimes, simply assuming uniform distributions or randomly generating the initial (nonzero) values are appropriate for the model. Then the initial values must be normalized by letting $\sum_{j,d}\hat{\pi}_{j,d}=1$, $\sum_{j\neq i,d}\hat{a}_{(i,h)(j,d)}=1$, and $\sum_{v_{k_1},\ldots,v_{k_d}}\hat{b}_{j,d}(v_{k_1:k_d})=1$.

However, the initial values of the model parameters can affect the trained model parameters in practice. Using the trained model parameters, the estimated states may have no intuitive sense on the physical process. For instance, a higher indexed state may correspond to a lower arrival rate of traffic. Therefore, it is more important to carefully set the initial values of the observation distributions compared with the initial values of the state transition probabilities and the duration distributions.

Example 4.5 Meaning States

Use Example 1.4. Based on empirical analysis of the workload data set, the total number of states, M, is about 20, and the maximum duration of state, D, is about 500. The initial values of λ_0 are simply assumed to be uniformly distributed, with the initial mean of each observation distribution corresponding to the index of the given state. For instance, let the initial value of $\mu_j=\max\{o_t\}\times j/M=3.7j$, for $j=1,\ldots,M$. This assumption makes the higher state corresponding to the higher arrival rate. Apply the estimation procedure to estimate the model parameters λ. The procedure converges to a fixed point of model parameters after about 20 iterations. The estimated arrival rate for each given state is listed as follows:

State:	*1*	*2*	*3*	*4*	*5*	*6*	*7*	*8*	*9*	*10*
Arrv rate:	*13*	*15.5*	*16*	*17.6*	*18*	*20.8*	*22.7*	*25.4*	*27.4*	*30.1*
State:	*11*	*12*	*13*	*14*	*15*	*16*	*17*	*18*	*19*	*20*
Arrv rate:	*31.8*	*35.6*	*37.6*	*37.9*	*40.1*	*43*	*44.7*	*47.8*	*58*	*60*

4.4.7 Termination Condition for the Estimation Procedure

It is a practical problem how to judge when the procedure of the model parameter estimation can be terminated. Usually, the procedure of the model parameter estimation is as follows: Set an initial model parameter set λ_0 and let $k=0$. Then for given model parameter set λ_k, estimate $\hat{\pi}_{j,d}$, $\hat{a}_{(i,h)(j,d)}$, and $\hat{b}_{j,d}(v_{k_1:k_d})$. Finally let $\hat{\lambda}_{k+1} = \{\hat{\pi}_{j,d}, \hat{a}_{(i,h)(j,d)}, \hat{b}_{j,d}(v_{k_1:k_d})\}$, $\lambda_{k+1} = \hat{\lambda}_{k+1}$, and $k = k + 1$.

In the estimation procedure, every time when λ_{k+1} is obtained, use $\sum_{j\in\mathbf{S}}\sum_{d\in\mathbf{D}}\alpha_T(j,d)$ to compute the likelihood $L(\lambda_{k+1}) = P[o_{1:T}|\lambda_{k+1}]$. If $[L(\lambda_{k+1}) - L(\lambda_k)]/L(\lambda_k) < \varepsilon$, then let $\lambda = \lambda_{k+1}$ and terminate the re-estimation procedure; otherwise go back to repeat the estimation procedure, where ε is a given small number used as the criteria of termination, such as $\varepsilon = 5\%$. Except the termination criteria $\varepsilon = 5\%$, it is often setting the maximum iteration number, say 20. Other termination criteria could be the relative difference of the model parameters, for example, $\sum_{i,h,j,d}\frac{|\hat{a}_{(i,h)(j,d)} - a_{(i,h)(j,d)}|}{MD} \leq \theta$.

CHAPTER 5

Conventional HSMMs*

This chapter discusses four conventional models that are often applied in the literature, with fewer parameters and lower computational complexity than the general model.

5.1 EXPLICIT DURATION HSMM

Ferguson (1980) was the first to consider the HSMM, which is called an "HMM with variable duration." Since then a number of studies have been reported on the subject (see, e.g., Mitchell and Jamieson, 1993; Yu and Kobayashi, 2003a, 2006, and references therein).

The explicit duration HSMM assumes that a state transition is independent of the duration of the previous state, that is, $a_{(i,h)(j,d)} = a_{i(j,d)}$, without self-transitions, that is, $a_{i(i,d)} = 0$. The state duration is assumed to be dependent on the current state and independent of the previous state. That is, state j will last for duration variable d according to the conditional probability $p_j(d)$, as defined in Eqn (2.2). Therefore, we have $a_{(i,h)(j,d)} = a_{ij}p_j(d)$ with $a_{ii} = 0$, for $i, j \in \mathbf{S}, d \in \mathbf{D}$, where $a_{ij} \equiv P[S_{[t} = j | S_{t-1]} = i]$ is the state transition probability from state i to state j. It also assumes the conditional independence of outputs as defined in Eqn (3.15). Due to all those independence assumptions, the explicit duration HSMM is one of the simplest models among all of the HSMMs. Therefore, it is the most popular HSMM in applications. Figure 5.1 shows the explicit duration hidden Markov model.

Replace $a_{(i,h)(j,d)}$ with $a_{ij}p_j(d)$, $b_{j,d}(o_{t+1:t+d})$ with $\prod_{\tau=t+1}^{t+d} b_j(o_\tau)$, and $\beta_t(j,d)$ with $\beta_t(j) \equiv P[o_{t+1:T} | S_{t]} = j, \lambda]$ in the general forward–backward formulas (2.6) and (2.7), and define $\alpha_t(j) \equiv P[S_{t]} = j, o_{1:t} | \lambda] = \sum_{d \in \mathbf{D}} \alpha_t(j,d)$. Then we readily obtain the forward–backward formulas for the explicit duration HSMM (Ferguson, 1980):

*To distinguish the conventional HSMMs from HMMs, we will call explicit duration HMM as "explicit duration HSMM," variable transition HMM as "variable transition HSMM," and residual time HMM as "residual time HSMM" in the rest of this book.

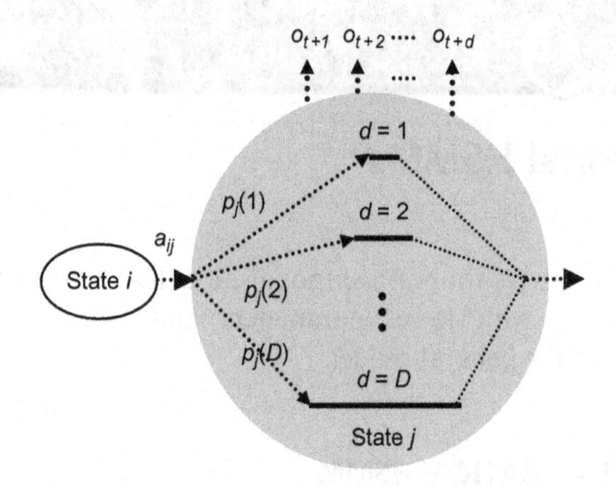

Figure 5.1 Explicit duration HSMM.
The dotted lines represent the instantaneous transitions with zero time, and the thick lines represent the sojourn time the state spends. After state i ends at the tth time unit, it transits to state $j \neq i$, according to the transition probability a_{ij}, and then selects its duration d according to the duration distribution $p_j(d)$. State j spends d time units and produces d observations $o_{t+1}, \ldots, o_{t+d}$ with emission probability $b_j(o_{t+1:t+d})$. When state j ends at the $(t+d)$th time unit, it transits to another state.

$$\alpha_t(j) \equiv P[S_{t]} = j, o_{1:t} | \lambda] = \sum_{d \in \mathbf{D}} \alpha^*_{t-d+1}(j) p_j(d) u_t(j,d), \tag{5.1}$$

$$\alpha^*_{t+1}(j) \equiv P[S_{[t+1} = j, o_{1:t} | \lambda] = \sum_{i \in \mathbf{S} \setminus \{j\}} \alpha_t(i) a_{ij}, \tag{5.2}$$

for $j \in \mathbf{S}$, $t = 1, \ldots, T$, and

$$\beta^*_{t+1}(j) \equiv P[o_{t+1:T} | S_{[t+1} = j, \lambda] = \sum_{d \in \mathbf{D}} p_j(d) u_{t+d}(j,d) \beta_{t+d}(j), \tag{5.3}$$

$$\beta_t(j) \equiv P[o_{t+1:T} | S_{t]} = j, \lambda] = \sum_{i \in \mathbf{S} \setminus \{j\}} a_{ji} \beta^*_{t+1}(i), \tag{5.4}$$

for $j \in \mathbf{S}$, $t = T-1, \ldots, 0$, where

$$u_t(j,d) \equiv \prod_{\tau = t-d+1}^{t} b_j(o_\tau). \tag{5.5}$$

The forward variable $\alpha_t(j)$ represents the joint probability that state j ends at t and the partial observation sequence is $o_{1:t}$, and $\alpha^*_{t+1}(j)$ the joint probability that state j starts at $t+1$ and the partial observation sequence is $o_{1:t}$. The backward variable $\beta_t(j)$ represents the conditional probability that given state i ending at t, the future observation sequence is $o_{t+1:T}$, and $\beta^*_{t+1}(j)$ the conditional probability

Algorithm 5.1 Forward–Backward Algorithm for the Explicit Duration HSMM

The Forward Algorithm (*based on* Eqns (5.5), (5.1), (5.2))

1. *For* $i = 1, \ldots, M$, *let* $\alpha^*_1(i) = \pi_i$ *and* $\alpha^*_\tau(i) = 0$ *for* $\tau < 0$, *where the boundary condition takes the simplified assumption, that is, the first state must start at* $t = 1$;
2. *For* $t = 1, \ldots, T$ {
   for $j = 1, \ldots, M$, *for* $d = 1, \ldots, D$ {
   $u_t(j, d) = \prod_{\tau=t-d+1}^{t} b_j(o_\tau)$;
   $\alpha_t(j) \leftarrow \alpha_t(j) + \alpha^*_{t-d+1}(j) p_j(d) u_t(j, d)$;
   }
   for $j = 1, \ldots, M$ { $\alpha^*_{t+1}(j) = \sum_{i \in \mathbf{S} \setminus \{j\}} \alpha_t(i) a_{ij}$; }
   }

The Backward Algorithm (*based on* Eqns (5.3), (5.4))

1. *For* $j = 1, \ldots, M$, *let* $\beta_T(j) = 1$ and $\beta_\tau(j) = 0$ *for* $\tau > T$, *where the boundary condition takes the simplified assumption, that is, the last state must end at* $t = T$;
2. *For* $t = T - 1, \ldots, 0$ {
   for $j = 1, \ldots, M$ { $\beta^*_{t+1}(j) = \sum_{d \in \mathbf{D}} p_j(d) u_{t+d}(j, d) \beta_{t+d}(j)$; }
   for $j = 1, \ldots, M$ { $\beta_t(j) = \sum_{i \in \mathbf{S} \setminus \{j\}} a_{ji} \beta^*_{t+1}(i)$; }
   }

that given state j starting at $t+1$, the future observation sequence is $o_{t+1:T}$.

The boundary conditions use the simplified assumption, that is, $\alpha^*_1(i) = P[S_{[1} = j | \lambda] = \pi_i$ and $\alpha^*_\tau(i) = 0$ for $\tau < 0$, and $\beta_T(i) = 1$ and $\beta_\tau(i) = 0$ for all $\tau > T, i \in \mathbf{S}$, where π_i is the initial distribution of state i.

The forward–backward algorithm for the explicit duration HSMM is shown in Algorithm 5.1.

5.1.1 Smoothed Probabilities

In this model, the probability of state i occurring with duration d, defined by Eqn (2.10), is

$$\eta_t(i, d) = \alpha^*_{t-d+1}(i) p_i(d) u_t(i, d) \beta_t(i),$$

the probability of transition from state i to state j, defined by Eqn (2.12), is

$$\xi_t(i, j) = \alpha_t(i) a_{ij} \beta^*_{t+1}(j),$$

and the probability of state i used to produce observation o_t, defined by Eqn (2.13) and derived by Eqn (2.14), is

$$\begin{aligned}\gamma_t(i) &= \gamma_{t-1}(i) + \alpha_t^*(i)\beta_t^*(i) - \alpha_{t-1}(i)\beta_{t-1}(i) \\ &= \pi_i\beta_1^*(i) + \sum_{\tau=2}^{t}\left[\alpha_\tau^*(i)\beta_\tau^*(i) - \alpha_{\tau-1}(i)\beta_{\tau-1}(i)\right],\end{aligned} \tag{5.6}$$

where according to the simplified assumption of the boundary conditions

$$\gamma_1(i) = P[S_1 = i, o_{1:T}|\lambda] = P[S_{[1} = i, o_{1:T}|\lambda] = \pi_i\beta_1^*(i)$$

and

$$\gamma_T(i) = P[S_T = i, o_{1:T}|\lambda] = P[S_{T]} = i, o_{1:T}|\lambda] = \alpha_T(i).$$

Obviously, Eqn (5.6) can be calculated backward in the backward procedure of the algorithm. If the last state is assumed to end after the last observation is obtained, the probability of being state i at time T with observations $o_{1:T}$ is (Cohen et al., 1997)

$$\sum_{t_1 \le T \le t_2} P[S_{[t_1,t_2]} = i, o_{1:T}|\lambda] = \sum_{d \in \mathbf{D}} \alpha^*_{T-d+1}(i)\left[\sum_{\tau=d}^{D} p_i(\tau)\right] u_T(i,d).$$

Using the smoothed probabilities, the set of formulas for ML re-estimation of model parameters can be obtained:

$$\hat{a}_{ij} = \frac{\sum_t \xi_t(i,j)}{\sum_j\sum_t \xi_t(i,j)}, \tag{5.7}$$

$$\hat{b}_i(v_k) = \frac{\sum_t \gamma_t(i)\cdot I(o_t = v_k)}{\sum_t \gamma_t(i)}, \tag{5.8}$$

$$\hat{p}_i(d) = \frac{\sum_t \eta_t(i,d)}{\sum_d\sum_t \eta_t(i,d)},$$

and

$$\hat{\pi}_i = \frac{\gamma_1(i)}{\sum_j \gamma_1(j)},$$

where the denominators are the normalization factors.

5.1.2 Computational Complexity

From Eqn (5.5), we can see that $u_t(j,d)$, for given $j \in \mathbf{S}$ and t, requires $d = 1,\ldots, D$ multiplications, respectively, that is, total $D(D+1)/2$ multiplications. Therefore, at each time t, over $j = 1,\ldots, M$, it requires $O(MD^2/2)$ multiplications. Meanwhile, at each time t, the forward–backward formulas (5.1)–(5.4) require extra $2DM$, M^2, $2DM$, and M^2 multiplications, respectively. Therefore, the total computational complexity of the explicit duration HSMM is $O((MD^2 + MD + M^2)T)$. Except $M^2 + MD + MK$ model parameters that are required to be stored, the forward variables $\alpha_t(j)$ and $\alpha_t^*(j)$ are needed to be stored so that the probabilities $\eta_t(j,d)$, $\xi_t(i,j)$, $\gamma_t(j)$, or the estimations $\hat{a}_{ij}$, $\hat{b}_j(v_k)$, $\hat{p}_j(d)$, can be computed in the backward procedure, where K is the total number of observable values that O_t can take. Therefore, the total storage requirement is $O(M^2 + MD + MK + MT)$.

Because the computational complexity is high, the explicit duration HSMM is not appropriate to be applied in some applications when D is large. To reduce this computational complexity, the key is to reduce the computational complexity of $u_t(j,d)$. Levinson (1986a) suggested a recursive method that can be used to calculate the product more efficiently, that is,

$$u_t(j,d) = \prod_{\tau=t-d+1}^{t} b_j(o_\tau) = u_t(j,d-1)\cdot b_j(o_{t-d+1}) \tag{5.9}$$

with $u_t(j,1) = b_j(o_t)$, which requires $O(MD)$ multiplications. This recursive method was also used by Mitchell et al. (1995). However, in their method D recursive steps must be performed at every t. Therefore, the total number of recursive steps required in their method increases by a factor of D compared with the Ferguson algorithm (Ferguson, 1980).

In fact, a better way to reduce both the computational complexity and the total number of recursive steps is letting

$$u_t(j,d) = u_{t-1}(j,d-1)\cdot b_j(o_t), \tag{5.10}$$

which can be implemented in a parallel manner and has no need to retrieve previous observation probabilities $b_j(o_{t-d+1})$. This idea was realized in a parallel implementation of the explicit duration HSMM for spoken language recognition on a hardware architecture by Mitchell et al. (1993). The computational load of $p_j(d) \cdot u_t(j, d)$ can also be reduced by approximation such as segmental beam pruning and duration pruning as proposed by Russell (2005). It shows that they can combine to give a 95% reduction in segment probability computations at a cost of a 3% increase in phone error rate.

5.2 VARIABLE TRANSITION HSMM

In this model, an HSMM is realized in the HMM framework, including the 2-vector HMM (Krishnamurthy et al., 1991), the duration-dependent state transition model (Vaseghi, 1991, 1995; Vaseghi and Conner, 1992), the inhomogeneous HMM (Ramesh and Wilpon, 1992), and the nonstationary HMM (Sin and Kim, 1995; Djuric and Chun, 1999, 2002). These approaches take the vector (i, d) as a HMM state, where i is one of the HSMM states and d sojourn time since entering the state. The explicit duration HSMM can also be expressed in this model by letting the triples (i, w, d) be HMM states, where d is a duration and w a counter, $1 \le w \le d$, which indicates the number of observations produced so far while in state i (Ferguson, 1980).

In addition to the state and its sojourn time, Pieczynski et al. (2002) added the observation as the third component. This makes it possible to generalize the model to the triplet Markov chain (Pieczynski et al., 2002; Pieczynski, 2005, 2007; Pieczynski and Desbouvries, 2005; Lanchantin and Pieczynski, 2004, 2005; Lanchantin et al., 2008; Lapuyade-Lahorgue and Pieczynski, 2006; Ait-el-Fquih and Desbouvries, 2005). The constraints among the three components are released in the triplet Markov chain model and the components are extended to be general processes. The price is loss of physical meaning in the sense of hidden semi-Markov process. One has to add some constraints back on the triplet Markov chain and re-define the meaning of the three processes when it is applied for the HSMM. The triplet Markov chain model can be further generalized to be a nonstationary fuzzy Markov chain by letting the underlying Markov chain be a fuzzy Markov random chain (Salzenstein et al., 2007). It can also be used to model a nonstationary hidden semi-Markov chain by introducing a fourth component. The fourth component takes its values in a finite set of states. Each of the states models a given set of parameters defining a given distribution of the other three components (Lapuyade-Lahorgue and Pieczynski, 2012).

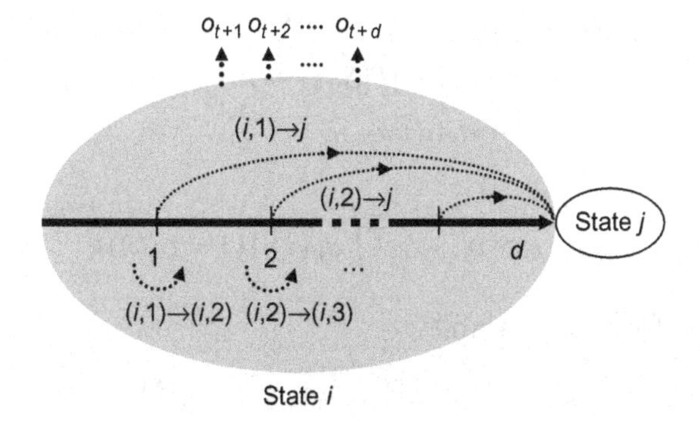

Figure 5.2 Variable transition HSMM.
The dotted lines represent the instantaneous transitions with zero time, and the thick lines represent the sojourn time the state spends. State i *starts at the* $(t+1)$*th time unit and continues to the* $(t+2)$*th unit with self-transition probability* $a_{ii}(1)$*. It will continue from* $(i, d-1)$ *to* (i, d) *with self-transition probability* $a_{ii}(d-1)$*. At any* $d \geq 1$*, state* i *can transit to another state* j *with transition probability* $a_{ij}(d)$*. During state* i*,* d *observations* $o_{t+1}, \ldots, o_{t+d}$ *are produced with emission probability* $b_i(o_{t+1:t+d})$*.*

Compared with the explicit duration HSMM, the variable transition HSMM assumes the state transition is dependent on the state duration, and hence it is more suitable for describing inhomogeneous or nonstationary hidden Markov processes. This makes it useful for some applications that cannot be modeled by a homogeneous process.

As shown in Figure 5.2, a state transition is allowed only for either $(i,d) \rightarrow (j,1)$, for $i \neq j$, or $(i,d) \rightarrow (i,d+1)$ for self-transitions. It assumes the conditional independence of outputs as given by Eqn (3.15). The boundary conditions use the simplified assumption. The state transition probability from state i to state j given that the sojourn time in state i at time t is d is defined by (Ramesh and Wilpon, 1992; Krishnamurthy et al., 1991)

$$a_{ij}(d) \equiv P[S_{t+1} = j | S_{[t-d+1:t} = i],$$

subject to $\sum_{j \in \mathbf{S}} a_{ij}(d) = 1$, for $i, j \in \mathbf{S}, d \in \mathbf{D}$, where the self-transition with probability $a_{ii}(d) > 0$ can occur. We note that $a_{ij}(d)$ is different from $a_{(i,d)j}$ defined by Eqn (2.1). The latter does not allow self-transition by assuming $a_{(i,d)i} = 0$. Since $S_t = j$ means state j either ends at t or continues to $t+1$, we have $P[S_t = j] = P[S_{t]} = j] + P[S_{t:t+1} = j]$. The probability that state j ends at t after entering the state for d is $P[S_{t]} = j | S_{[t-d+1:t} = j] = 1 - a_{jj}(d)$. Then, for $i \neq j$,

$$a_{ij}(d) = P[S_{t]} = i, S_{t+1} = j | S_{[t-d+1:t} = i]$$
$$= (1 - a_{ii}(d))P[S_{[t+1} = j | S_{[t-d+1:t]} = i]$$
$$= (1 - a_{ii}(d))a_{(i,d)j}$$

and

$$a_{(i,d)(j,h)} = a_{(i,d)j}\left(\prod_{\tau=1}^{h-1} a_{jj}(\tau)\right)(1 - a_{jj}(h)).$$

Define the forward variable

$$\widehat{\alpha}_t(j,d) \equiv P[S_{[t-d+1:t} = j, o_{1:t} | \lambda]$$

representing the joint probability that the sojourn time in state j at time t is d and the partial observation sequence is $o_{1:t}$, and the backward variable

$$\widehat{\beta}_t(j,d) \equiv P[o_{t+1:T} | S_{[t-d+1:t} = j, \lambda]$$

representing the conditional probability that the future observation sequence is $o_{t+1:T}$ given that the sojourn time in state j at time t is d. We have the following relationships:

$$\alpha_t(j,d) = P[S_{[t-d+1:t]} = j, o_{1:t} | \lambda] = \widehat{\alpha}_t(j,d)(1 - a_{jj}(d)),$$
$$\beta_t(j,d) = \sum_i P[S_{[t+1} = i, o_{t+1:T} | S_{[t-d+1:t]} = j, \lambda]$$
$$= \sum_i a_{(j,d)i} P[o_{t+1:T} | S_{[t+1} = i, \lambda]$$
$$= \sum_i a_{(j,d)i} b_i(o_{t+1}) \widehat{\beta}_{t+1}(i,1),$$

and

$$\widehat{\beta}_t(j,d) = P[S_{t]} = j, o_{t+1:T} | S_{[t-d+1:t} = j, \lambda]$$
$$+ P[S_{t+1} = j, o_{t+1:T} | S_{[t-d+1:t} = j, \lambda]$$
$$= (1 - a_{jj}(d))\beta_t(j,d) + a_{jj}(d) b_j(o_{t+1}) \widehat{\beta}_{t+1}(j, d+1).$$

Applying these equations or deriving directly from the definitions of $\widehat{\alpha}_t(j,d)$ and $\widehat{\beta}_t(j,d)$, the general forward–backward formulas (2.6) and (2.7) are reduced to those of the variable transition HSMM:

$$\widehat{\alpha}_t(j,d) = \begin{cases} \sum_{h \in \mathbf{D}} \sum_{i \in \mathbf{S} \setminus \{j\}} \widehat{\alpha}_{t-1}(i,h) a_{ij}(h) b_j(o_t), & d = 1 \\ \widehat{\alpha}_{t-1}(j, d-1) a_{jj}(d-1) b_j(o_t), & d > 1 \end{cases} \tag{5.11}$$

for $j \in \mathbf{S}$, $d \in \mathbf{D}$, $t = 2, \ldots, T$, and

$$\widehat{\beta}_t(j,d) = \sum_{i \in \mathbf{S} \setminus \{j\}} a_{ji}(d)\widehat{\beta}_{t+1}(i,1)b_i(o_{t+1}) + a_{jj}(d)\widehat{\beta}_{t+1}(j,d+1)b_j(o_{t+1}) \tag{5.12}$$

for $j \in \mathbf{S}, d \in \mathbf{D}, t = T-1, \ldots, 1$, where the conditional independence of outputs as given by Eqn (3.15) is assumed.

The boundary conditions are $\widehat{\alpha}_1(j,1) = \pi_j b_j(o_1)$, $\widehat{\alpha}_1(j,d) = 0$ for $d > 1$ and $\widehat{\beta}_T(j,d) = 1$, for $j \in \mathbf{S}, d \in \mathbf{D}$. Similar to the forward recursion formula, a Viterbi algorithm for the inhomogeneous HMM can be readily obtained by replacing the sum $\sum_{h \in \mathbf{D}} \sum_{i \in \mathbf{S} \setminus \{j\}}$ of formula (5.11) with the maximum operations $\max_{h \in \mathbf{D}} \max_{i \in \mathbf{S} \setminus \{j\}}$, as done by Ramesh and Wilpon (1992) and Deng and Aksmanovic (1997).

The forward–backward algorithm for the variable transition HSMM is shown in Algorithm 5.2.

Algorithm 5.2 Forward–Backward Algorithm for the Variable Transition HSMM

The Forward Algorithm (*based on* Eqn (5.11))

1. *For* $j = 1, \ldots, M$, *let* $\widehat{\alpha}_1(j,1) = \pi_j b_j(o_1)$, $\widehat{\alpha}_1(j,d) = 0$ *for* $d > 1$, *where the boundary condition takes the simplified assumption, that is, the first state must start at* $t = 1$;
2. *For* $t = 2, \ldots, T$, *for* $j = 1, \ldots, M$ {

   $\widehat{\alpha}_t(j,1) = \sum_{i \in \mathbf{S} \setminus \{j\}} \widehat{\alpha}_{t-1}(i,1) a_{ij}(1) b_j(o_t)$;

   for $d = 2, \ldots, D$ {

   $\widehat{\alpha}_t(j,1) \leftarrow \widehat{\alpha}_t(j,1) + \sum_{i \in \mathbf{S} \setminus \{j\}} \widehat{\alpha}_{t-1}(i,d) a_{ij}(d) b_j(o_t)$;

   $\widehat{\alpha}_t(j,d) = \widehat{\alpha}_{t-1}(j,d-1) a_{jj}(d-1) b_j(o_t)$;

   }

   }

The Backward Algorithm (*based on* Eqn (5.12))

1. *For* $j = 1, \ldots, M$ *and* $d = 1, \ldots, D$, *let* $\widehat{\beta}_T(j,d) = 1$, *where the boundary condition takes the simplified assumption, that is, the last state must end at* $t = T$;
2. *For* $t = T-1, \ldots, 1$, *for* $j = 1, \ldots, M$, *for* $d = 1, \ldots, D$ {

   $\widehat{\beta}_t(j,d) = \sum_{i \in \mathbf{S} \setminus \{j\}} a_{ji}(d)\widehat{\beta}_{t+1}(i,1)b_i(o_{t+1}) + a_{jj}(d)\widehat{\beta}_{t+1}(j,d+1)b_j(o_{t+1})$;

   }

5.2.1 Smoothed Probabilities

The joint probability that state (i,d) transits to state j at time $t+1$ and observation sequence takes $o_{1:T}$ given the model parameters is

$$\begin{aligned}\xi_t(i,d;j) &\equiv P[S_{[t-d+1:t} = i, S_{t+1} = j, o_{1:T}|\lambda] \\ &= \bar{\alpha}_t(i,d)a_{ij}(d)b_j(o_{t+1})\bar{\beta}_{t+1}(j,1), \quad i \neq j\end{aligned}$$

and

$$\xi_t(i,d;i) = \bar{\alpha}_t(i,d)a_{ii}(d)b_i(o_{t+1})\bar{\beta}_{t+1}(i,d+1), \quad d \leq D-1.$$

Then the smoothed probability of state transition from state (i,d) to state $j \neq i$ at time $t+1$ given the model and the observation sequence is $\xi_t(i,d;j)/P[o_{1:T}|\lambda]$. The smoothed probability of being in state i for duration d at time t given the model and the observation sequence, as defined in Eqn (2.10), is $\eta_t(i,d)/P[o_{1:T}|\lambda]$, and we have

$$\eta_t(i,d) = \sum_{j \in \mathbf{S}\setminus\{i\}} \xi_t(i,d;j)$$

We also have $\xi_t(i,j) = \sum_d \xi_t(i,d;j)$ and $\gamma_t(j) = \gamma_{t+1}(j) + \sum_{i \in \mathbf{S}\setminus\{j\}}[\xi_t(j,i) - \xi_t(i,j)]$, as yielded by Eqn (2.14).

Then $\gamma_1(i)$ can be used to estimate the initial probabilities $\hat{\pi}_i$, $\sum_t \gamma_t(j) I(o_t = v_k)$ for the observation probabilities $\hat{b}_j(v_k)$, and $\sum_t \xi_t(i,d;j)$ for the transition probabilities $\hat{a}_{ij}(d)$.

5.2.2 Computational Complexity

Though the super state space of the pairwise process (i,d) is $\mathbf{S} \times \mathbf{D}$ in the order of MD, its real computational amount is lower than $(MD)^2T$, where M is the number of HSMM states and D the maximum duration of an HSMM state. From the forward formula (5.11), we can see that computing $\bar{\alpha}_t(j,1)$ for given t and j requires $2(M-1)$ D multiplications, and computing $\bar{\alpha}_t(j,d)$, for all $d>1$, requires $2(D-1)$ multiplications. Therefore, computing the forward variables over all j and t requires $(2MD-2)MT$ multiplications. From the backward formula (5.12), we can see that computing the backward variables requires $2M(MD)T$ multiplications. Therefore, the total computational complexity is $O(M^2DT)$. Except $M^2D + MK$ model parameters required to be stored, the forward variables $\bar{\alpha}_t(j,d)$, over all t, j, and d, are required to be stored for computing the smoothed probabilities and estimation of the model parameters along with the backward procedure, where K is the total number of observable values. Therefore, the total storage requirement is $O(M^2D + MK + MDT)$. Compared with $O((M^2 + MD + MD^2)T)$ of the explicit duration HSMM, the computational complexity of the variable transition HSMM is higher when the order of the state space is higher, and is lower when the maximum length of the state durations is smaller. However, its space complexity is definitely higher than $O(M^2 + MD + MK + MT)$ of the explicit duration HSMM.

5.3 VARIABLE-TRANSITION AND EXPLICIT-DURATION COMBINED HSMM

Though the variable transition HSMM and the explicit duration HSMM have different assumptions for their models, their model parameters can be expressed by each other. Since

$$\begin{aligned} p_i(d)a_{ij} &= P[S_{t-d+1:t]} = i, S_{[t+1} = j | S_{[t-d+1} = i] \\ &= \left(\prod_{\tau=1}^{d-1} a_{ii}(\tau)\right) \cdot a_{ij}(d), \end{aligned}$$

the model parameters of the explicit duration HSMM can be expressed by those of the variable transition HSMM as

$$\begin{aligned} p_i(d) &= P[S_{t-d+1:t]} = i | S_{[t-d+1} = i] \\ &= \prod_{\tau=1}^{d-1} a_{ii}(\tau) \cdot [1 - a_{ii}(d)], \end{aligned}$$

and $a_{ij} = a_{ij}(d)/[1 - a_{ii}(d)]$. Reversely, since $a_{ii}(D) = 0$ and

$$\begin{aligned} \sum_{h \geq d} p_i(h) &= \sum_{h=d}^{D-1} \prod_{\tau=1}^{h-1} a_{ii}(\tau) \cdot [1 - a_{ii}(h)] + \prod_{\tau=1}^{D-1} a_{ii}(\tau) \\ &= \sum_{h=d}^{D-2} \prod_{\tau=1}^{h-1} a_{ii}(\tau) \cdot [1 - a_{ii}(h)] + \prod_{\tau=1}^{D-2} a_{ii}(\tau) \\ &= \ldots = \prod_{\tau=1}^{d-1} a_{ii}(\tau), \end{aligned}$$

the model parameters of the variable transition HSMM can be expressed by those of the explicit duration HSMM (Djuric and Chun, 2002; Azimi et al., 2005):

$$a_{ii}(d) = \frac{\sum_{h \geq d+1} p_i(h)}{\sum_{h \geq d} p_i(h)}, \tag{5.13}$$

$$a_{ij}(d) = [1 - a_{ii}(d)]a_{ij} = \frac{p_i(d)}{\sum_{h \geq d} p_i(h)} a_{ij}, \tag{5.14}$$

for $i \neq j$ and $i,j \in \mathbf{S}$, $d \in \mathbf{D}$.

These two models can be combined by assuming $a_{(i,h)(j,d)} = a_{(i,h)j} p_j(d)$ (Marhasev et al., 2006). The forward–backward formulas (2.6) and (2.7) then become those of the variable-transition and explicit-duration combined HSMM:

$$\alpha_{t+d}(j,d) = \left(\sum_{i \in \mathbf{S} \backslash \{j\}} \sum_{h \in \mathbf{D}} \alpha_t(i,h) a_{(i,h)j} \right) p_j(d) b_{j,d}(o_{t+1:t+d})$$

and

$$\beta_t(j,d) = \sum_{i \in \mathbf{S} \backslash \{j\}} a_{(j,d)i} \left(\sum_{h \in \mathbf{D}} p_i(h) b_{i,h}(o_{t+1:t+h}) \beta_{t+h}(i,h) \right).$$

Denote $X_t(j) = \left(\sum_{i \in \mathbf{S} \backslash \{j\}} \sum_{h \in \mathbf{D}} \alpha_t(i,h) a_{(i,h)j} \right)$ and $Y_t(i) = \left(\sum_{h \in \mathbf{D}} p_i(h) b_{i,h}(o_{t+1:t+h}) \beta_{t+h}(i,h) \right)$, for $i,j \in \mathbf{S}$. Then $\alpha_{t+d}(j,d) = X_t(j) p_j(d) b_{j,d}(o_{t+1:t+d})$ and $\beta_t(j,d) = \sum_{i \in \mathbf{S} \backslash \{j\}} a_{(j,d)i} Y_t(i)$.

Obviously, the computational complexity for $\{X_t(j) : j\}$ is $O(M^2 D)$, for $\{\alpha_{t+d}(j,d) : j,d\}$ is $O(MD)$, for $\{Y_t(i) : i\}$ is $O(MD)$, and for $\{\beta_t(j,d) : j,d\}$ is $O(M^2 D)$. Therefore, the computational complexity of this model is $O((M^2 D + MD)T)$, where the computation amount for $b_{j,d}(o_{t+1:t+d})$ is not included.

5.4 RESIDUAL TIME HSMM

The residual time HSMM (Yu and Kobayashi, 2003a) assumes a state transition is either $(i,1) \rightarrow (j,\tau)$ for $i \neq j$ or $(i,\tau) \rightarrow (i,\tau-1)$ for a self-transition with $\tau > 1$, where τ is the residual time of state i. The state transition probabilities are assumed to be independent of the

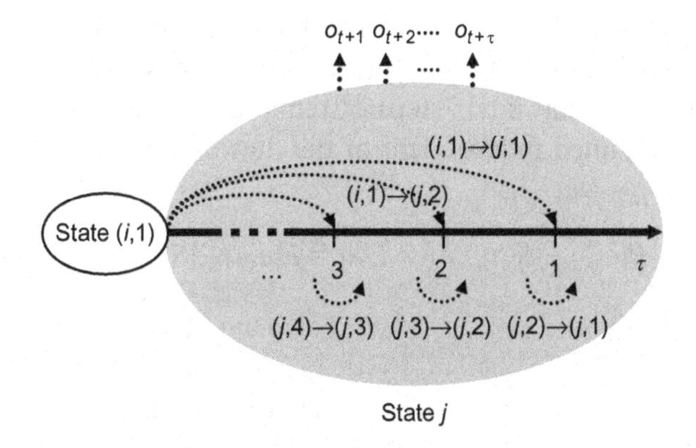

Figure 5.3 Residual time HSMM.
The dotted lines represent the instantaneous transitions with zero time, and the thick lines represent the sojourn time the state spends. The residual time of a state is denoted by τ in the figure. After state i ends at the tth time unit (with residual time 1), it transits to state j with a selected duration/residual time τ according to the transition probability $a_{i(j,\tau)}$. State j with residual time τ will continue to reduce its residual time to $\tau - 1$ until 1. Then it ends at the $(t + \tau)$th time unit and transits to another state. During state j, τ observations $o_{t+1}, \ldots, o_{t+\tau}$ are produced with emission probability $b_j(o_{t+1:t+\tau})$.

duration of the previous state. The residual time HSMM also assumes the conditional independence of outputs as yielded by Eqn (3.15). The boundary conditions use the simplified assumption. The model is shown in Figure 5.3 and its DBN is shown in Figure 3.1(b). Therefore, this model is useful in the application areas where the residual time that the current state will stay in the future is of interest. It is in contrast to the variable transition HSMM which is interested in the sojourn time that the current state has been stayed in the past.

As defined by Eqn (2.3), the state transition probability from state i to state j that will have residual time τ is $a_{i(j,\tau)} \equiv P[S_{[t:t+\tau-1]} = j | S_{t-1]} = i]$ for $i \neq j$, with $\sum_{j \in \mathbf{S} \setminus \{i\}} \sum_{\tau \in \mathbf{D}} a_{i(j,\tau)} = 1$. The self-transition probability from (i, τ) to $(i, \tau - 1)$ is $P[S_{t+1:t+\tau-1]} = i | S_{t:t+\tau-1]} = i] = 1$, for $\tau > 1$.

Define the *forward variables* and *backward variables* by (Yu and Kobayashi, 2003a)

$$\breve{\alpha}_t(i, \tau) \equiv P[S_{t:t+\tau-1]} = i, o_{1:t} | \lambda]$$

and

$$\breve{\beta}_t(i, \tau) \equiv P[o_{t+1:T} | S_{t:t+\tau-1]} = i, \lambda].$$

The forward variable $\breve{\alpha}_t(i, \tau)$ is the joint probability that the partial observation sequence is $o_{1:t}$ and the current state i will stay for the next

τ steps and then end at $t+\tau-1$. The backward variable $\breve{\beta}_t(i,\tau)$ is the conditional probability that the future observations will be $o_{t+1:T}$ given the current state i that has τ steps of remaining time. Because the state transition is assumed independent of the duration of the previous state, we have $a_{(i,d)(j,\tau)} = a_{i(j,\tau)}$,

$$\beta_t(i,d) = P[o_{t+1:T}|S_{[t-d+1:t]} = i, \lambda] = P[o_{t+1:T}|S_{t]} = i, \lambda] = \breve{\beta}_t(i,1),$$

for any $d \in \mathbf{D}$, and

$$\breve{\beta}_t(i,\tau) = \breve{\beta}_{t+\tau-1}(i,1) \prod_{k=t+1}^{t+\tau-1} b_i(o_k),$$

where the conditional independence of outputs is assumed, as given by Eqn (3.15). We also have the following relationships:

$$\sum_{d\geq\tau} \alpha_t(i,d) = P[S_{t-\tau+1:t]} = i, o_{1:t}|\lambda] = \breve{\alpha}_{t-\tau+1}(i,\tau) \prod_{k=t-\tau+2}^{t} b_i(o_k)$$

with $\sum_{d\geq 1} \alpha_t(i,d) = \breve{\alpha}_t(i,1)$, and

$$\begin{aligned}\alpha_t(j,d) &= P[S_{[t-d+1:t]} = j, o_{1:t}|\lambda] \\ &= \sum_{i\neq j} P[S_{t-d]} = i, S_{[t-d+1:t]} = j, o_{1:t}|\lambda] \\ &= \sum_{i\neq j} \breve{\alpha}_{t-d}(i,1) a_{i(j,d)} \prod_{k=t-d+1}^{t} b_j(o_k).\end{aligned}$$

Applying those relationships or deriving from the definitions of $\breve{\alpha}_t(i,\tau)$ and $\breve{\beta}_t(i,\tau)$, the general forward–backward formulas (2.6) and (2.7) are reduced to those of the residual time HSMM:

$$\breve{\alpha}_t(i,\tau) = \breve{\alpha}_{t-1}(i,\tau+1) b_i(o_t) + \sum_{j\in\mathbf{S}\setminus\{i\}} \breve{\alpha}_{t-1}(j,1) a_{j(i,\tau)} b_i(o_t), \tag{5.15}$$

for $i \in \mathbf{S}$, $\tau \in \mathbf{D}$, $t = 1,\ldots,T$, and

$$\breve{\beta}_t(i,\tau) = b_i(o_{t+1}) \breve{\beta}_{t+1}(i,\tau-1), \quad \tau > 1, \tag{5.16}$$

$$\breve{\beta}_t(i,1) = \sum_{j\in\mathbf{S}\setminus\{i\}} \sum_{\tau\geq 1} a_{i(j,\tau)} b_j(o_{t+1}) \breve{\beta}_{t+1}(j,\tau), \tag{5.17}$$

for $i \in \mathbf{S}$, $\tau \in \mathbf{D}$, $t = T-1,\ldots,1$. The boundary conditions are $\breve{\alpha}_0(i,1) = \pi_i$, $\breve{\alpha}_0(i,\tau) = 0$ for $\tau > 1$, and $\breve{\beta}_T(i,1) = 1$, $\breve{\beta}_T(i,\tau) = 0$ for $\tau > 1$.

The computational complexity involved in the residual time HSMM can be reduced significantly if the state duration is assumed to be independent of the previous state. In this case, we have $a_{i(j,\tau)} = a_{ij}p_j(\tau)$. From the definition of $\alpha^*_{t+1}(i)$ given by Eqn (5.2) and the definition of $\beta^*_{t+1}(j)$ given by Eqn (5.3),

$$\alpha^*_t(i) = P[S_{[t} = i, o_{1:t-1}|\lambda] = \sum_{j\in\mathbf{S}\backslash\{i\}} \breve{\alpha}_{t-1}(j,1)a_{ji} \tag{5.18}$$

and

$$\beta^*_{t+1}(j) = P[o_{t+1:T}|S_{[t+1} = j] = b_j(o_{t+1})\sum_{\tau\in\mathbf{D}} p_j(\tau)\breve{\beta}_{t+1}(j,\tau). \tag{5.19}$$

Then the forward formula (5.15) and the backward formula (5.17) are reduced to (Yu and Kobayashi, 2003a)

$$\breve{\alpha}_t(i,\tau) = \breve{\alpha}_{t-1}(i,\tau+1)b_i(o_t) + \alpha^*_t(i)p_i(\tau)b_i(o_t) \tag{5.20}$$

and

$$\breve{\beta}_t(i,1) = \sum_{j\in\mathbf{S}\backslash\{i\}} a_{ij}\beta^*_{t+1}(j). \tag{5.21}$$

The forward–backward algorithm for the residual time HSMM is shown in Algorithm 5.3. The simplified algorithm with reduced computational complexity is presented in Algorithm 5.4.

5.4.1 Smoothed Probabilities

The smoothed probabilities can be obtained as follows:

$$\begin{aligned}\xi_t(i;j,d) &\equiv P[S_{t]} = i, S_{[t+1:t+d]} = j, o_{1:T}|\lambda] \\ &= \breve{\alpha}_t(i,1)a_{i(j,d)}b_j(o_{t+1})\breve{\beta}_{t+1}(j,d),\end{aligned} \tag{5.22}$$

for $i \neq j$,

$$\eta_{t+d}(i,d) \equiv P[S_{[t+1:t+d]} = i, o_{1:T}|\lambda] = \sum_{i\in\mathbf{S}\backslash\{j\}} \xi_t(i;j,d),$$

$$\xi_t(i,j) = \sum_d \xi_t(i;j,d), \tag{5.23}$$

and

$$\gamma_t(i) \equiv P[S_t = i, o_{1:T}|\lambda] = \sum_d \breve{\alpha}_t(i,d)\breve{\beta}_t(i,d),$$

Algorithm 5.3 Forward–Backward Algorithm for the Residual Time HSMM

The Forward Algorithm (*based on* Eqn (5.15))

1. *For* $i = 1, \ldots, M$, *let* $\breve{\alpha}_0(i,1) = \pi_i$, $\breve{\alpha}_0(i,\tau) = 0$ *for* $\tau > 1$, *where the boundary condition takes the simplified assumption, that is, the first state must start at* $t = 1$;
2. *For* $t = 1, \ldots, T$, *for* $i = 1, \ldots, M$, *for* $\tau = 1, \ldots, D$ {
   $\breve{\alpha}_t(i,\tau) = \breve{\alpha}_{t-1}(i,\tau+1)b_i(o_t) + \sum_{j\in \mathbf{S}\backslash\{i\}} \breve{\alpha}_{t-1}(j,1)a_{j(i,\tau)}b_i(o_t)$;
   }

The Backward Algorithm (based on Eqns (5.17), (5.16))

1. *For* $i = 1, \ldots, M$, *let* $\breve{\beta}_T(i,1) = 1$, $\breve{\beta}_T(i,\tau) = 0$ *for* $\tau > 1$, *where the boundary condition takes the simplified assumption, that is, the last state must end at* $t = T$;
2. *For* $t = T-1, \ldots, 1$, *for* $i = 1, \ldots, M$ {
   $\breve{\beta}_t(i,1) = \sum_{j\in \mathbf{S}\backslash\{i\}} a_{i(j,1)}b_j(o_{t+1})\breve{\beta}_{t+1}(j,1)$
   for $\tau = 2, \ldots, D$ {
   $\breve{\beta}_t(i,1) \leftarrow \breve{\beta}_t(i,1) + \sum_{j\in \mathbf{S}\backslash\{i\}} a_{i(j,\tau)}b_j(o_{t+1})\breve{\beta}_{t+1}(j,\tau)$;
   $\breve{\beta}_t(i,\tau) = b_i(o_{t+1})\breve{\beta}_{t+1}(i,\tau-1)$;
   }
   }

with the initial condition

$$\gamma_T(i) = \sum_{d\geq 1} \breve{\alpha}_T(i,d), \tag{5.24}$$

where $\gamma_t(i)$ can also be computed using Eqn (2.14).

The MATLAB code for computing the smoothed probabilities, $\gamma_t(j)$ and $\xi_t(i,j)$, and estimating the states of the residual time HSMM is quite simple as shown in Algorithm 5.4. Its initialization code for the algorithm can be found from the website http://sist.sysu.edu.cn/~syu/. With slight changes, this code can be used to estimate the model parameters.

5.4.2 Computational Complexity

Now we consider the forward recursion given by Eqns (5.18) and (5.20), and the backward recursion by Eqns (5.19), (5.21) and (5.16). We can see that computing the forward variables $\alpha_t^*(i)$ for all i requires $O(M^2)$ steps, and $\breve{\alpha}_t(i,\tau)$ for all i and τ requires extra $O(MD)$ steps.

Algorithm 5.4 MATLAB Code for State Estimation of the Residual Time HSMM

% $PI = [\pi_i]_{M\times 1}$, $B = [b_i(v_k)]_{M\times K}$, $P = [p_i(\tau)]_{M\times D}$, $A = [a_{ij}]_{M\times M}$ with $a_{ii} = 0$

%--- Forward Algorithm --- % based on Eqns (5.18), (5.20)	
ALPHA = [PI, zeros(M, D-1)];	% $\breve{\alpha}_0(i,1) = \pi_i$, others are zeros
for t = 1:T	
x = repmat(((A'*ALPHA(:,1)).*B(:,O(t))),1,D).*P;	%$x = \left(\sum_{j\in \mathbf{S}\setminus\{i\}} \breve{\alpha}_{t-1}(j,1)a_{ji}\right)p_i(\tau)b_i(o_t)$
w = ALPHA(:,2:D).*repmat(B(:,O(t)),1,D-1);	% $w = \breve{\alpha}_{t-1}(i,\tau+1)b_i(o_t)$
ALPHA = [w,zeros(M,1)] + x;	% $\breve{\alpha}_t(i,\tau) = [w,0] + x$
c(t) = 1/sum(ALPHA(:));	% c(t) the scaling factor
ALPHA = ALPHA.*c(t);	% scaled $\breve{\alpha}_t(i,\tau)$
ALPHAx(:,t) = ALPHA(:,1);	% record the forward results
end	
log-likelihood = −sum(log(c));	% log-likelihood: $\ln P[o_{1:T}\|\lambda]$
%--- Backward Algorithm --- % backward by Eqns (5.19), (5.21), (5.16) % state estimation by Eqns (5.24), (2.15)	
BETA = [ones(M,1),zeros(M,D-1)];	% $\breve{\beta}_T(i,1) = 1$, others are zeros
GAMMA = sum(ALPHA,2);	% $\gamma_T(i) = \sum_{d\geq 1}\breve{\alpha}_T(i,d)$
[u,S_est(T)] = max(GAMMA);	% $\hat{s}_T = \arg\max_i \gamma_T(i)$
for t = (T-1):-1:1	
y = B(:,O(t+1)).*c(t+1);	% to scale $\breve{\beta}_t(i,\tau)$by scaling $b_j(o_{t+1})$
z = y.*(sum((P.*BETA),2));	% $z_j = b_j(o_{t+1})(\sum_{\tau\geq 1} p_j(\tau)\breve{\beta}_{t+1}(j,\tau))$
BETA(:,2:D) = repmat(y,1,D-1).*BETA(:,1:D-1);	% $\breve{\beta}_t(i,\tau) = b_i(o_{t+1})\breve{\beta}_{t+1}(i,\tau-1)$
BETA(:,1) = A*z;	% $\breve{\beta}_t(i,1) = \sum_{j\in \mathbf{S}\setminus\{i\}} a_{ij}z_j$
XI = (ALPHAx(:,t)*z').*A;	% $\xi_t(i,j) = \breve{\alpha}_t(i,1)a_{ij}z_j$
GAMMA = GAMMA + sum(XI,2)-sum(XI',2);	% $\gamma_t(j) = \gamma_{t+1}(j) + \sum_i[\xi_t(j,i) - \xi_t(i,j)]$
[u,S_est(t)] = max(GAMMA);	% $\hat{s}_t = \arg\max_i \gamma_t(i)$
end	

Similarly, computing the backward variables $\beta^*_{t+1}(j)$ for all j requires $O(MD)$ steps, and $\breve{\beta}_t(i,1)$ for all i requires extra $O(M^2)$ steps. Hence, the total number of computation steps for evaluating the forward and backward variables is $O((MD + M^2)T)$. This computational complexity is much lower than those of the explicit duration HSMM and the variable transition HSMM.

The backward variables $\breve{\beta}_t(i,\tau)$ and the probabilities $\eta_t(i,\tau)$, $\xi_t(i,j)$, and $\gamma_t(i)$ do not need to be stored for estimation of the model parameters $\{p_i(\tau), a_{ij}, b_i(o_t), \pi_i\}$. Only the forward variables $\breve{\alpha}_t(i,1)$ and $\alpha_t^*(i)$ for all i and t need to be stored, with the storage requirement of $O(MT)$. Therefore, the storage requirement for the residual time HSMM is $O(M^2 + MD + MK + MT)$, similar to the explicit duration HSMM.

CHAPTER 6

Various Duration Distributions

This chapter presents the most popular parametric distributions of state duration.

6.1 EXPONENTIAL FAMILY DISTRIBUTION OF DURATION

The choice of distribution family for the state duration is central to the use of the HSMM (Duong et al., 2006). The state duration is usually modeled using Poisson (Russell and Moore, 1985), Gaussian (Ariki and Jack, 1989; Yoma and Sanchez, 2002), and gamma distributions (Levinson, 1986a; Yoma and Sanchez, 2002). All these distributions belong to the exponential family (Levinson, 1986a; Mitchell and Jamieson, 1993). Some other complex duration models can be found in Ostendorf et al. (1996). A discussion on the capacity and complexity of duration modeling techniques can be found in Johnson (2005).

The probability density function (pdf) or probability mass function (pmf) for the duration of state j belonging to the exponential family can be expressed as (Mitchell and Jamieson, 1993)

$$p_j(d;\boldsymbol{\theta}_j) = \frac{1}{B(\boldsymbol{\theta}_j)}\xi(d)\exp\left(-\sum_{p=1}^{P}\theta_{j,p}S_p(d)\right), \tag{6.1}$$

where P is the number of natural parameters, $\theta_{j,p}$ is the pth natural parameter for state j, $\boldsymbol{\theta}_j = (\theta_{j,1},\ldots,\theta_{j,P})$ is the set of parameters for state j, $S_p(d)$ and $\xi(d)$ are sufficient statistic, and $B(\boldsymbol{\theta}_j)$ is a normalizing term satisfying

$$B(\boldsymbol{\theta}_j) = \int_0^{\infty}\xi(x)\exp\left(-\sum_{p=1}^{P}\theta_{j,p}S_p(x)\right)dx.$$

From definitions (2.10) and (2.11), we can see that $\eta_{t+d}(j,d) = \sum_{i\neq j}\sum_{h=1}^{\min\{D,t\}}\xi_t(i,h;j,d)$. By substituting $a_{(i,h)(j,d)} = a_{(i,h)j}p_j(d)$

in Eqn (3.9), the model parameters for the duration distributions can be found by

$$\max_{\theta} \sum_{(j,d)} \left(\sum_{t=1}^{T-d} \frac{\eta_{t+d}(j,d)}{P[o_{1:T}|\lambda]} \right) \log \frac{p_j(d;\theta)}{p_j(d)},$$

and in considering Eqn (3.12) and $\theta = (\theta_1, \ldots, \theta_M)$, the new parameters for the duration distribution of state j can be found by maximizing the following function (Ferguson, 1980)

$$X_P(\theta_j) \equiv \sum_{d=1}^{D} \hat{p}_j(d) \log p_j(d;\theta_j),$$

where $\hat{p}_j(d)$ is the nonparametric pmf estimated by the re-estimation formula (3.12), and $\sum_{d=1}^{D} \hat{p}_j(d) = 1$. Since the exponential family is log-concave, the global maximum can be found by setting the derivative equal to zero, yielding the maximum likelihood equations

$$\frac{\partial}{\partial \theta_{j,p}} X_P(\theta_j) = \sum_{d=1}^{D} \hat{p}_j(d) \left[-\frac{\partial \log B(\theta_j)}{\partial \theta_{j,p}} - S_p(d) \right] = 0,$$

where

$$-\frac{\partial \log B(\theta_j)}{\partial \theta_{j,p}} = \frac{1}{B(\theta_j)} \int_0^{\infty} \xi(x) \exp\left(-\sum_{p=1}^{P} \theta_{j,p} S_p(x) \right) S_p(x) dx = E(S_p(x)|\theta_j)$$

is the expected value taken with respect to the exponential family member. Therefore, the new duration parameters can be determined by letting

$$E(S_p(x)|\theta_j) = \sum_{d=1}^{D} \hat{p}_j(d) S_p(d), \quad p = 1, \ldots, P. \tag{6.2}$$

Example 6.1 State Duration with Gaussian Distributions

Since $\exp\left(-\frac{(d-\mu)^2}{2\sigma^2}\right) = \exp\left(-\frac{\mu^2}{2\sigma^2}\right) \cdot \exp\left(-\frac{d^2 - 2d\mu}{2\sigma^2}\right)$, we have $P = 2$, $S_1(d) = d$, $\theta_{j,1} = -\mu/\sigma^2$, $S_2(d) = d^2$, $\theta_{j,2} = 1/2\sigma^2$, and $\xi(d) = 1$ for the Gaussian distribution. Then the new parameters for the Gaussian distribution of state duration can be estimated by letting $\hat{\mu} = \sum_{d=1}^{D} \hat{p}_j(d) d$ and $\hat{\sigma}^2 = \sum_{d=1}^{D} \hat{p}_j(d) d^2 - \hat{\mu}^2$.

From Eqn (6.1), we can see that $\sum_{d=1}^{D} p_j(d) \neq 1$. That is, the pmf sampled or truncated from the corresponding pdf is an approximation. Hence, before applying the pmf into the forward–backward algorithms of HSMMs, it is necessary to introduce a normalizing factor c_j such that $\sum_{d=1}^{D} c_j p_j(d) = 1$. In this sense, instead of getting parameters for the pdf by Eqn (6.2), solving for the pmf directly may yield a higher likelihood (Mitchell and Jamieson, 1993). Let

$$B(\boldsymbol{\theta}_j) = \sum_{d=1}^{D} \xi(d) \exp\left(-\sum_{p=1}^{P} \theta_{j,p} S_p(d)\right).$$

Then $E(S_p(x)|\boldsymbol{\theta}_j) = \sum_{d=1}^{D} p_j(d;\boldsymbol{\theta}_j) S_p(d)$ and the accurate parameters $\hat{\boldsymbol{\theta}}_j$ can be found by solving the following equations:

$$\sum_{d=1}^{D} p_j(d;\boldsymbol{\theta}_j) S_p(d) = \sum_{d=1}^{D} \hat{p}_j(d) S_p(d), \quad p = 1, \ldots, P.$$

From these equations, we can see that numerical solutions may be required to find parameters for the pmf.

6.2 DISCRETE COXIAN DISTRIBUTION OF DURATION

When the state durations are discrete Coxian distributions, the HSMM can be formulated as an HMM and benefit from the HMM formulae and methods. This will result in that the computational amount required for the corresponding HMM is independent of the length of state durations. In contrast, the conventional HSMMs may encounter huge computational amount in summing over all possible length of durations when the state durations have infinite support.

Denote a discrete Coxian distribution by $Cox(\mu, \theta)$, where $\mu = (\mu_1, \ldots, \mu_N)$ and $\theta = (\theta_1, \ldots, \theta_N)$ are parameters (Duong et al., 2005a, 2005b). A left-to-right Markov chain with N states (phases/stages) is used to describe $Cox(\mu, \theta)$. Each phase $n \in \{1, \ldots, N\}$ has the transition probability θ_n, $0 < \theta_n \leq 1$, to the next phase $n+1$, with the self-transition probability $A_{nn} = 1 - \theta_n$ and the geometric distribution of duration $X_n \sim Geom(\theta_n)$. If the left-to-right Markov chain starts from phase n, then $X_n + \cdots + X_N$ is the duration of the left-to-right Markov chain. The probability that the left-to-right Markov chain starts from phase n is μ_n, $0 \leq \mu_n \leq 1$, $\sum_n \mu_n = 1$.

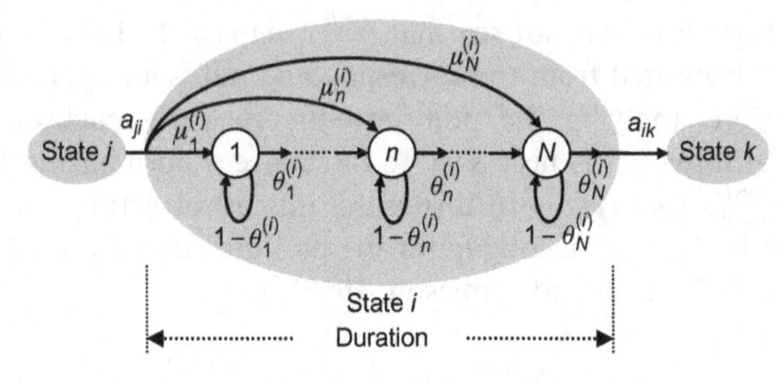

Figure 6.1 HSMM state starts from any phase n *and always ends at phase* N.
State j *transits to state* i *with transition probability* a_{ji}. *After entering state* i, *it randomly selects phase* n *with probability* $\mu_n^{(i)}$. *Then starting from phase* n, *it transits to itself with probability* $(1-\theta_n^{(i)})$ *or the next phase* n + 1 *with probability* $\theta_n^{(i)}$ *until phase* N. *Phase* N *may transit to itself with probability* $(1-\theta_N^{(i)})$ *or departure phase* N *and state* i *with probability* $\theta_N^{(i)}$. *Then transits to the next state* k *with transition probability* a_{ik}. *The duration of state* i *is the sum of the durations of phase* n *to phase* N.

If a transition from phase n to any phase $m \geq n$ is allowed, with $A_{nm} \geq 0$, for $m \geq n$ and $n, m \in \{1, \ldots, N\}$, the left-to-right Markov chain is in fact a series–parallel network of geometric processes. Based on Coxian theory, the overall duration pdf of the series–parallel network can construct any discrete pdf with rational *z-transform* (Wang, 1994; Bonafonte et al., 1996; Wang et al., 1996).

A discrete Coxian distribution of duration for a conventional HSMM assumes that the duration distribution $p_i(d)$ for given state $i \in S$ is $Cox(\mu^{(i)}, \theta^{(i)})$ with $d = X_n^{(i)} + \cdots + X_N^{(i)}$, where $\mu^{(i)} = (\mu_1^{(i)}, \ldots, \mu_N^{(i)})$ and $\theta^{(i)} = (\theta_1^{(i)}, \ldots, \theta_N^{(i)})$ are the Coxian parameter set for the HSMM state $i \in S$, and $X_n^{(i)} \sim Geom(\theta_n^{(i)})$, as shown in Figure 6.1. When $\theta_n^{(i)} = 1$ for all n, that is, without self-transition of phases, it reduces to the conventional duration distribution of HSMMs with $X_n \equiv 1$, $p_i(d) = \mu_{N-d+1}^{(i)}$, and $D = N$.

Example 6.2 State duration with Coxian Distributions

For a Coxian distribution with two phases, the state duration distribution is

$$p_i(d) = c_1(A_{2,2}^{(i)})^{d-1} - c_2(A_{1,1}^{(i)})^{d-1}$$

where $A_{n,n}^{(i)} = 1 - \theta_n^{(i)}$, for $n = 1,2$, $c_1 = \frac{\mu_1^{(i)}\theta_1^{(i)}\theta_2^{(i)}}{\theta_1^{(i)} - \theta_2^{(i)}} + \mu_2^{(i)}\theta_2^{(i)}$, and $c_2 = \frac{\mu_1^{(i)}\theta_1^{(i)}\theta_2^{(i)}}{\theta_1^{(i)} - \theta_2^{(i)}}$. For different selection of parameters $\mu^{(i)} = (\mu_1^{(i)}, \mu_2^{(i)})$ and $\theta^{(i)} = (\theta_1^{(i)}, \theta_2^{(i)})$, the duration of state i can have different distributions other than geometric distributions. For example, let $\theta_1^{(i)} = 0.2$, $\theta_2^{(i)} = 0.1$, $\mu_1^{(i)} = 0.7$, and $\mu_2^{(i)} = 0.3$,

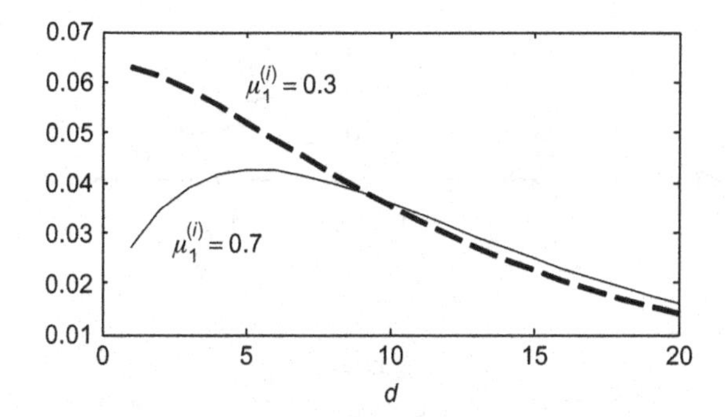

Figure 6.2 Coxian Distributions with two phases.
The solid line represents the Coxian distribution of two phases with model parameters $\theta_1^{(i)} = 0.2$, $\theta_2^{(i)} = 0.1$, $\mu_1^{(i)} = 0.7$, and $\mu_2^{(i)} = 0.3$. The dashed line represents the Coxian distribution of two phases with model parameters $\theta_1^{(i)} = 0.2$, $\theta_2^{(i)} = 0.1$, $\mu_1^{(i)} = 0.3$, and $\mu_2^{(i)} = 0.7$. They are different from the geometric distributions.

the distribution is the solid line of Figure 6.2. While let $\mu_1^{(i)} = 0.3$ and $\mu_2^{(i)} = 0.7$, the distribution is the dashed line of Figure 6.2.

Obviously, (i, n) can be considered an HMM state, where $i \in \mathbf{S}$ is the state of the HSMM and $n \in \{1, \ldots, N\}$ is the phase of the Coxian distribution of the state. Therefore, the traditional forward–backward algorithm for HMM can be applied for the model parameter re-estimation and state sequence estimation.

An HMM state (i, n) can transit to (i, n) with self-transition probability $A_{nn}^{(i)} = 1 - \theta_n^{(i)}$ for any n, to $(i, n+1)$ with probability $A_{n,n+1}^{(i)} = \theta_n^{(i)}$ for $n < N$, and to (j, n') with probability $a_{ij}\mu_{n'}^{(j)}$ for $n = N$, where $A_{nm}^{(i)}$, $\theta_n^{(i)}$, and $\mu_n^{(i)}$ are parameters for state i. The computational complexity is $O(M^2NT)$. As an extension of such HMM (Russell and Cook, 1987), a transition from phase n to any phase m is allowed, with $A_{nm} \geq 0$, for $n \neq m$ and $n, m \in \{1, \ldots, N\}$.

An equivalent model is assuming that the left-to-right Markov chain of each HSMM state always starts from phase 1 and ends at any phase n, as shown in Figure 6.3, where $A_{n,e}^{(j)}$ is the probability that state j ends at phase n, $A_{n,n+1}^{(j)}$ the transition probability from phase (j, n) to $(j, n+1)$, and $A_{n,n}^{(j)}$ the self-transition probability, with $A_{n,e}^{(j)} + A_{n,n+1}^{(j)} + A_{n,n}^{(j)} = 1$ and $A_{N,N+1}^{(j)} = 0$.

Let $\alpha_t[(i, n), d]$ be the forward variable at time t denoting the probability that the state is i, the phase is n, and the duration of state i is d

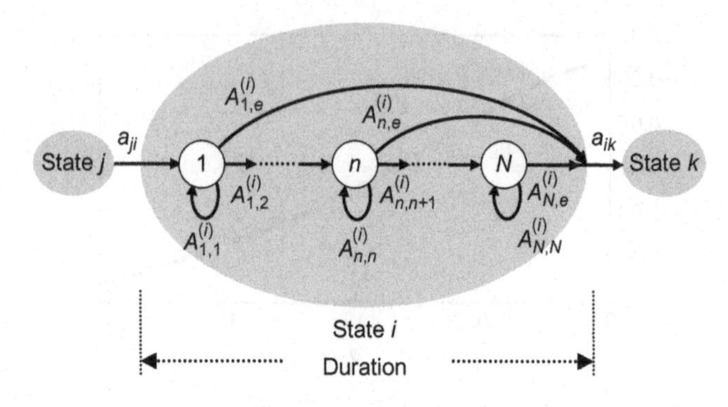

Figure 6.3 HSMM state always starts from phase 1 and ends at any phase n.
After entering state i, *it enters phase* 1. *In phase* n, *it may transit to itself with probability* $A^{(i)}_{n,n}$, *or next phase* n + 1 *with probability* $A^{(i)}_{n,n+1}$, *or out of state* i *with probability* $A^{(i)}_{n,e}$. *Phase* N *cannot transit to other phase, and so the transition probability* $A^{(i)}_{N,N+1} = 0$. *After state* i *ends, it transits to the next state* k *with transition probability* a_{ik}. *The duration of state* i *is the sum of the durations of phase* 1 *to* n *if state* i *ends at phase* n.

(Wang, 1994; Wang et al., 1996; Sitaram and Sreenivas, 1997). Then the forward algorithm can be expressed by

$$\alpha_t[(i,n),d] = \begin{cases} \sum_{j\in\mathbf{S}}\sum_{m=1}^{N}\sum_{h\in\mathbf{D}} \alpha_{t-1}[(j,m),h]A^{(j)}_{m,e}a_{ji}b_{i,1}(o_t), & n=1, d=1 \\ \alpha_{t-1}[(i,1),d-1]A^{(i)}_{1,1}b_{i,1}(o_t), & n=1, d>1 \\ \sum_{m=n-1}^{n} \alpha_{t-1}[(i,m),d-1]A^{(i)}_{mn}b_{i,n}(o_t), & d\geq n>1 \\ 0, & d<n \end{cases} \tag{6.3}$$

where $b_{i,n}(o_t)$ is the observation probability at phase (i,n). The Viterbi version of this forward formula can be straightforward by replacing Σ with *max* as shown by Kwon and Un (1995) and Peng et al. (2000).

This model can be reduced to a simpler model (Langrock and Zucchini, 2011) by letting the self-transition probabilities of phases be zeros except the last phase N, that is, $A^{(i)}_{n,n} = 0$ for $n < N$, and $A^{(i)}_{N,N} > 0$. In other words, the duration of the last phase has geometric distribution. Denote the phase transition probabilities $a_i(d) = A^{(i)}_{d,d+1}$, $1 - a_i(d) = A^{(i)}_{d,e}$, for $d < N$, and $a_i(d) = a_i(N) = A^{(i)}_{N,N}$, $1 - a_i(d) = 1 - a_i(N) = A^{(i)}_{N,e}$, for $d \geq N$. In this case, if $n < N$, the length of the state duration is equal to the number of phases that have been passed, that is, $\alpha_t[(i,n),d] = 0$ for $n \neq d$ and $n < N$. Therefore, we can denote $\alpha'_t(i,d) = \alpha_t[(i,d),d]$.

for $d<N$. Similarly denote $\alpha'_t(i,d)=\alpha_t[(i,N),d]$, for $d\geq N$. Then Eqn (6.3) is reduced to

$$\alpha'_t(i,d)=\begin{cases}\sum_{j\in\mathbf{S}}\sum_{h\in\mathbf{D}}\alpha'_{t-1}(j,h)(1-a_j(h))a_{ji}b_{i,1}(o_t), & d=1\\ \alpha'_{t-1}(i,d-1)a_i(d-1)b_{i,d}(o_t), & d>1\end{cases}. \tag{6.4}$$

From the second equation of (6.4), it is easy to derive that, for $1<d\leq N$,

$$\alpha'_t(i,d)=\alpha'_{t-d+1}(i,1)\prod_{h=1}^{d-1}(a_i(h)b_{i,h+1}(o_{t-d+h+1}))$$

and, for $d>N$,

$$\alpha'_t(i,d)=\alpha'_{t-d+1}(i,1)\left(\prod_{h=1}^{N-1}a_i(h)\right)a_i(N)^{d-N}\left(\prod_{h=1}^{d-1}b_{i,h+1}(o_{t-d+h+1})\right).$$

By substituting these two equations into the first equation of (6.4), it can be seen that the state duration distribution is $p_i(d)=(1-a_i(d))\prod_{h=1}^{d-1}a_i(h)$ for $d\leq N$, and $p_i(d)=p_i(N)a_i(N)^{d-N}$ for $d>N$. This means that for any pmf $\{p_i(d):d=1,\ldots,\infty\}$ of HSMM state duration, we can find corresponding parameters $\{a_i(d):d=1,\ldots,N\}$ for the left-to-right Markov chain to exactly represent the first part, $\{p_i(d):d\leq N\}$, of the pmf by letting $a_i(1)=1-p_i(1)$, and $a_i(d)=1-p_i(d)/\prod_{h=1}^{d-1}a_i(h)$, for $d=2,\ldots,N$. Its higher duration part for $\{p_i(d):d>N\}$ is approximated using the geometric distribution $p_i(d)\approx p_i(N)a_i(N)^{d-N}$.

6.3 DURATION DISTRIBUTIONS FOR VITERBI HSMM ALGORITHMS

Similar to HSMMs, Viterbi HSMM algorithms, as in Burshtein (1995, 1996) and Yoma et al. (1998, 2001), can use parametric distributions to describe the state duration. For some special distributions, such as concave monotonic distributions presented in the following example, the computational amount in finding the maximum overall possible state durations can be reduced.

For the explicit duration HSMM, the Viterbi HSMM algorithm given by Eqn (2.29) becomes

$$\begin{aligned}\delta_t(j)&=\max_{S_{1:t-1}}P[S_{1:t-1},S_{t]}=j,o_{1:t}|\lambda]\\&=\max_{(t-d,i)\in Q(j)}\left\{\delta_{t-d}(i)a_{ij}p_j(d)\prod_{t'=t-d+1}^{t}b_j(o_{t'})\right\},\end{aligned} \tag{6.5}$$

for $1 \leq t \leq T, j \in \mathbf{S}$, where $Q(j) = \{(t-d, i) : d \in \mathbf{D}, i \in \mathbf{S} \backslash \{j\}\}$. Tweed et al. (2005) found that if $p_j(d)$ is *concave monotonic*, that is,

$$C_1 p_j(d_1) \leq C_2 p_j(d_2) \Rightarrow C_1 p_j(d_1 + h) \leq C_2 p_j(d_2 + h),$$

for $d_1 > d_2$, $h > 0$ and constants C_1 and C_2, then from

$$\delta_{t-d_1}(i_1) a_{i_1 j} p_j(d_1) \prod_{\tau=t-d_1+1}^{t} b_j(o_\tau) \leq \delta_{t-d_2}(i_2) a_{i_2 j} p_j(d_2) \prod_{\tau=t-d_2+1}^{t} b_j(o_\tau) \quad (6.6)$$

we know that at time $t+h$

$$\delta_{t-d_1}(i_1) a_{i_1 j} p_j(d_1 + h) \prod_{\tau=t-d_1+1}^{t+h} b_j(o_\tau) \leq \delta_{t-d_2}(i_2) a_{i_2 j} p_j(d_2 + h) \prod_{\tau=t-d_2+1}^{t+h} b_j(o_\tau).$$

This means that for given state j if the longer segmentation (of length d_1, starting at earlier time $t - d_1$) has a lower probability, then it will always have a lower probability as both segmentations are further extended (Tweed et al., 2005). This fact can be used to reduce the number of items in $Q(j)$ of Eqn (6.5). That is, if Eqn (6.6) is satisfied, then $\delta_{t-d_1}(i_1)$ can never give optimal solutions in the future and the item $(t - d_1, i_1)$ can be removed from the set, $Q(j)$, for state j. The newer items (t, i), for all $i \in \mathbf{S}$, are pushed into $Q(j)$ after $\delta_t(j)$ for all $j \in \mathbf{S}$ are determined using Eqn (6.5). Meanwhile the oldest items $(t - D, i)$, for all $i \in \mathbf{S}$, are removed from $Q(j)$. Therefore, we usually have $|Q(j)| \ll (M-1)D$, that is, the computational amount of Eqn (6.5) is reduced.

CHAPTER 7

Various Observation Distributions

This chapter presents typical observation distributions that are often used in modeling applications.

7.1 TYPICAL PARAMETRIC DISTRIBUTIONS OF OBSERVATIONS

The observation variable $O_t \in V = \{v_1, \ldots, v_K\}$ is usually assumed as a discrete variable with finite alphabet $|V| = K$. In some applications, however, a parametric distribution with possibly infinite support may be required or preferred. For example, to model the arrival rate of packets in a network, the probability distribution $b_j(O_t)$ may be represented as a Poisson distribution with $O_t \in \{0, 1, \ldots, \infty\}$. In some other applications, the observation variable O_t may be treated as a continuous variable. For example, to model received signal plus noise, a continuous Gaussian random variable is often used. Using the parametric distributions, the number of model parameters for the observations can be reduced substantially (Ferguson, 1980).

Based on Eqn (3.10), the model parameters of parametric observation distributions can be estimated with increasing likelihood by finding

$$\max_{\varphi} \sum_{(j,d)} \sum_{t=d}^{T} \eta_t(j, d) \log b_{j,d}(o_{t-d+1:t}, \varphi), \quad (7.1)$$

where $\eta_t(j, d) = P[S_{[t-d+1:t]} = j, o_{1:T} | \lambda]$ is the joint probability that state j lasts from time $t-d+1$ to t with duration d and the observation sequence is $o_{1:T}$ given the model parameters, and $b_{j,d}(\cdot, \varphi)$ is the probability density/mass function of observation for given state j and duration d, with a set of model parameters φ to be determined. The probabilities $\eta_t(j, d)$ are determined by Eqn (2.10) using a forward–backward algorithm of HSMM. If the observations are assumed conditional independent for given state, that is, $b_{j,d}(o_{t-d+1:t}, \varphi) = \prod_{\tau=t-d+1}^{t} b_j(o_\tau, \varphi)$, Eqn (7.1) becomes

$$\max_{\varphi} \sum_{j} \sum_{t=1}^{T} \gamma_t(j) \log b_j(o_t, \varphi), \quad (7.2)$$

where $\gamma_t(j) = P[S_t = j, o_{1:T}|\lambda]$ is the joint probability that the state at time t is j and the observation sequence is $o_{1:T}$ given the model parameters, which is determined by Eqn (2.13). Since $\sum_{t=1}^{T}\gamma_t(j)I(o_t = v_k)/P[o_{1:T}|\lambda]$ is the expected number of times that observable v_k occurred, $\sum_j\sum_{t=1}^{T}\log b_j(o_t, \varphi)^{\frac{\gamma_t(j)}{P[o_{1:T}|\lambda]}}$ can be considered the log-likelihood function, where $I(o_t = v_k)$ is an indicator function. Then the model parameters of the observation distribution of state j can be estimated by

$$\max_{\varphi_j}\sum_{t=1}^{T}\gamma_t(j)\log b_j(o_t, \varphi_j), \tag{7.3}$$

for $j\in\mathbf{S}$. For example, if the pmf $b_j(O_t)$, for $O_t\in\{0, 1, \ldots, \infty\}$, is Poisson with mean μ_j, that is, $b_j(k) = \mu_j^k e^{-\mu_j}/k!$, then the parameter μ_j can be estimated by

$$\hat{\mu}_j = \frac{\sum_{t=1}^{T}\gamma_t(j)o_t}{\sum_{t=1}^{T}\gamma_t(j)} \tag{7.4}$$

or, equivalently, $\hat{\mu}_j = \Sigma_k\hat{b}_j(k)k$, where $\hat{b}_j(k)$ is the expectation of $b_j(k)$ given by Eqn (3.13). A similar result can be obtained by directly maximizing the likelihood function $P[o_{1:T}|\lambda]$ (Levinson, 1986a).

For the continuous random variable O_t, the observation distribution $b_j(O_t)$ of state j is often represented by a Gaussian distribution. The mean of the Gaussian distribution can be estimated by Eqn (7.4), and the variance σ^2 can be estimated by

$$\hat{\sigma}_j^2 = \frac{\sum_{t=1}^{T}\gamma_t(j)(o_t-\hat{\mu}_j)^2}{\sum_{t=1}^{T}\gamma_t(j)}. \tag{7.5}$$

If the observation distribution belongs to the exponential family, it will have a simple formula similar to Eqn (6.2) for estimating its parameters, that is,

$$E(S_l(x)|\varphi_j) = \frac{\sum_{t=1}^{T}\gamma_t(j)S_l(o_t)}{\sum_{t=1}^{T}\gamma_t(j)}, \quad l = 1, \ldots, L,$$

where $b_j(x,\varphi_j)=\frac{1}{B(\varphi_j)}\xi(x)\exp\left(-\sum_{l=1}^{L}\varphi_{j,l}S_l(x)\right)$, and $\varphi_j=(\varphi_{j,1},\ldots,\varphi_{j,L})$ are the model parameters.

7.2 A MIXTURE OF DISTRIBUTIONS OF OBSERVATIONS

More generally, a mixture of distributions, such as a mixture of Gaussian distributions (Oura et al., 2006), is assumed as an observation distribution that can be any form other than the exponential family distributions.

7.2.1 Countable Mixture of Distributions

An observation distribution consists of a number of distributions, and which distribution will be applied is a random variable. Specifically, for a given state j, the pdf that state j produces an observation v_k is written as a countable mixture of distributions (Huang, 1992)

$$b_j(v_k)=\sum_n p_{jn}f_{jn}(v_k), \tag{7.6}$$

where $f_{jn}(v_k)$ is the nth pdf of observations, and p_{jn} is the probability that the nth distribution $f_{jn}(v_k)$ is selected as the observation distribution of v_k, with $\sum_n p_{jn}=1$. Equivalently, we can extend state j into countable substates, and let $R_t=(j,n)$ denote that the process is in the n'th substate of state j at time t. Accordingly, define $p_{j,n}>0$ as the probability that the process is in the n'th substate of state j, as shown in Figure 7.1. When the process is in substate (j, n), the observation distribution is $f_{jn}(v_k)$.

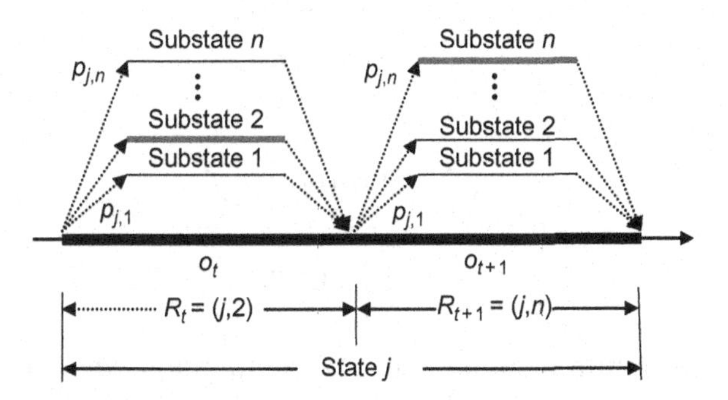

Figure 7.1 A state is extended to multiple substates.
After entering state j, it randomly selects a substate n to enter with probability $p_{j,n}$. In the figure, it actually selects substate 2 to enter. In substate 2, it produces an observation o_t according to the PDF $f_{j2}(o_t)$. Then it randomly selects the next substate n with probability $p_{j,n}$. In substate n, it produces an observation o_{t+1} according to the pdf $f_{jn}(o_{t+1})$. Suppose the duration of state j is 2. Then after substate n ends at time t + 1, state j also ends and transits to the next state at time t + 2.

By replacing $S_{1:T}$ with $R_{1:T}$ and following the derivation of Eqns (3.1)–(3.10), the model parameters of the observation distributions can be estimated by

$$\max_{\{p_{j,n},\varphi_{j,n};j,n\}} \sum_{j}\sum_{n}\sum_{t=1}^{T} \gamma_t(j,n)\log(p_{j,n}f_{j,n}(o_t,\varphi_{j,n})), \tag{7.7}$$

where $\gamma_t(j,n) = P[R_t = (j,n), o_{1:T}|\lambda]$, and $f_{j,n}(o_t,\varphi_{j,n}) > 0$ is assumed. By solving Eqn (7.7) s.t. $\sum_n p_{jn} = 1$, the model parameters for substate (j,n) can be estimated as follows:

$$\hat{p}_{j,n} = \frac{\sum_{t=1}^{T}\gamma_t(j,n)}{\sum_{n}\sum_{t=1}^{T}\gamma_t(j,n)}, \tag{7.8}$$

$$\sum_{t=1}^{T}\gamma_t(j,n)\frac{1}{f_{j,n}(o_t,\varphi_{j,n})}\frac{\partial f_{j,n}(o_t,\varphi_{j,n})}{\partial\varphi_{j,n}^{(l)}} = 0, \tag{7.9}$$

for $l = 1, \ldots, L$, where $\varphi_{j,n} = (\varphi_{j,n}^{(1)}, \ldots, \varphi_{j,n}^{(L)})$ is the set of model parameters of $f_{j,n}(o_t,\varphi_{j,n})$, and is to be determined by solving Eqn (7.9).

If $f_{j,n}(o_t,\varphi_{j,n})$ is a Gaussian distribution, Eqn (7.9) can be solved by letting

$$\hat{\mu}_{j,n} = \frac{\sum_{t=1}^{T}\gamma_t(j,n)o_t}{\sum_{t=1}^{T}\gamma_t(j,n)}, \tag{7.10}$$

$$\hat{\sigma}_{j,n}^2 = \frac{\sum_{t=1}^{T}\gamma_t(j,n)(o_t - \hat{\mu}_j)^2}{\sum_{t=1}^{T}\gamma_t(j,n)}. \tag{7.11}$$

Similar estimation formulas can also be found in Xie et al. (2012).

Now we consider how to compute $\gamma_t(j,n) = P[R_t = (j,n), o_{1:T}|\lambda]$ for given set of model parameters λ. It is the joint probability that the observation sequence is $o_{1:T}$, the state is j at time t and the substate is n, given the model parameters. In other words, the observation at time

t is produced by the substate n of state j, while the other observations other than time t are produced by all possible substates of states. Therefore, the joint probabilities of observations $o_{\tau-d+1:\tau}$, for $\tau-d+1\leq t\leq\tau$, is $f_{j,n}(o_t)\prod_{\substack{k=\tau-d+1\\k\neq t}}^{\tau}b_j(o_k)$, in considering that Eqn (7.7) implicitly assumes that the observations are conditionally independent. Then from Eqns (2.13), (2.10), and (2.6),

$$\begin{aligned}\gamma_t(j,n) &= \sum_{\substack{\tau,d:\\ \tau\geq t\geq\tau-d+1}} P[S_{[\tau-d+1:\tau]}=j, R_t=(j,n), o_{1:T}|\lambda]\\ &= \sum_{\substack{\tau,d:\\ \tau\geq t\geq\tau-d+1}} \sum_{i\neq j;h}\alpha_{\tau-d}(i,h)a_{(i,h)(j,d)}\prod_{\substack{k=\tau-d+1\\ k\neq t}}^{\tau} b'_j(o_k)p'_{j,n}f'_{j,n}(o_t)\beta_\tau(j,d)\\ &= \gamma_t(j)\frac{p'_{j,n}f'_{j,n}(o_t)}{b'_j(o_t)},\end{aligned} \tag{7.12}$$

where $p'_{j,n}\in\lambda$ are given model parameters and $f'_{jn}(o_t)$ and $b'_j(o_t)=\sum_n p'_{jn}f'_{jn}(o_t)$ are given observation distributions with given model parameters. Therefore, after the probabilities $\gamma_t(j)=P[S_t=j,o_{1:T}|\lambda]$ are calculated by Eqn (2.13) using a forward–backward algorithm of HSMM, $\gamma_t(j,n)$ can be determined by Eqn (7.12).

7.2.2 Uncountable Mixture of Distributions

When the observation O_t is assumed to have a distribution with an unknown parameter distributed as another distribution and then the parameter is marginalized out, the compounded observation distribution can be considered an uncountable mixture of distributions. Suppose $b_j(O_t)$ is compounded by an uncountable mixture of distributions $f_j(O_t;\nu)$. Then it can be expressed as

$$b_j(O_t)=\int_0^\infty p_j(\nu)f_j(O_t;\nu)d\nu.$$

For example, when $p_j(\nu)=G_j(\nu;\theta_j,\theta_j)$ is a Gamma distribution with the shape and scale parameters θ_j, and $f_j(O_t;\nu)=N_j\left(O_t;\mu_j,\Sigma_j/\nu\right)$ is a Gaussian distribution with mean μ_j and variable covariance Σ_j/ν, the compounded observation distribution $b_j(O_t)$ is another form of Student's t-distribution.

To estimate the model parameters, state j is thought of having infinite number of continuous substates (j, ν). The probability that the process is in substate (j, ν) of state j is $p_j(\nu)\Delta\nu$. Following a similar procedure in deriving Eqn (7.12), Eqn (7.7) becomes

$$\max_{\{\theta_j, \varphi_j : j\}} \sum_j \sum_{t=1}^{T} \frac{\gamma_t(j)}{b_j(o_t)} \int_0^\infty p_j(\nu) f_j(o_t; \nu) \log(p_j(\nu; \theta_j) f_j(o_t; \varphi_j, \nu)) d\nu,$$

where $o_{1:T}$ are the observations, $b_j(o_t)$, $p_j(\nu)$, and $f_j(o_t; \nu)$ are given model parameters and distributions, and θ_j and φ_j are the model parameters to be estimated. When $p_j(\nu)$ is a Gamma distribution and $f_j(O_t; \nu) = N_j\left(O_t; \mu_j, \Sigma_j/\nu\right)$ is a Gaussian distribution, the term $\log\left(G_j(\nu; \theta_j, \theta_j) N_j\left(o_t; \mu_j, \Sigma_j/\nu\right)\right)$ will contain ν and $\log(\nu)$. Thus it is required to compute the following expectations:

$$E(\nu) = \int_0^\infty p_j(\nu) f_j(o_t; \nu) \nu d\nu$$

$$E(\log \nu) = \int_0^\infty p_j(\nu) f_j(o_t; \nu) \log(\nu) d\nu,$$

before the model parameters θ_j, μ_j, and Σ_j can be estimated by partial derivative with respect to the parameters.

Because the Student's t-distribution does not belong to the exponential family distribution, a mixture of Student's t-distributions of Eqn (7.6) does not have the estimation formulae as simple as the mixture of Gaussian distributions. However, since a Student's t-distribution can be thought of being generated from an infinite mixture of Gaussian distribution with variable covariance distributed as gamma, we can extend state j into countable substates and each substate into a continuous sub-substate ν. Then the process being in the ν'th sub-substate of the n'th substate of state j at time t can be denoted as $R_t = (j, n, \nu)$, and Eqn (7.7) becomes

$$\max_{\{p_{j,n}, \mu_{j,n}, \Sigma_{j,n}, \theta_{j,n} : j,n\}} \sum_{j,n} \sum_{t=1}^{T} \frac{\gamma_t(j)}{b_j(o_t)} \int_0^\infty p'_{j,n} G_{j,n}(\nu; \theta'_{j,n}, \theta'_{j,n}) N_{j,n}\left(o_t; \mu'_{j,n}, \frac{\Sigma'_{j,n}}{\nu}\right)$$
$$\cdot \log\left(p_{j,n} G_{j,n}(\nu; \theta_{j,n}, \theta_{j,n}) N_{j,n}\left(o_t; \mu_{j,n}, \frac{\Sigma_{j,n}}{\nu}\right)\right) d\nu,$$

where o_t, $p'_{j,n}$, $\theta'_{j,n}$, $\mu'_{j,n}$, and $\Sigma'_{j,n}$ are given, $N_{j,n}(o_t; \mu_{j,n}, \Sigma_{j,n}/\nu)$ is a Gaussian distribution with mean $\mu_{j,n}$ and covariance $\Sigma_{j,n}/\nu$, and $G_{j,n}(\nu; \theta_{j,n}, \theta_{j,n})$ is a Gamma probability density function with the shape and scale parameters $\theta_{j,n}$. Now, the parameters $p_{j,n}$ can be readily estimated by Eqn (7.8), and $\theta_{j,n}$, $\mu_{j,n}$, and $\Sigma_{j,n}$ can be estimated after computing the expectations of $E(\nu)$ and $E(\log(\nu))$. Similar estimation formulae can be found in Ding and Shah (2010).

7.3 MULTISPACE PROBABILITY DISTRIBUTIONS

It is assumed that observation space consists of multiple subspaces, which can be overlapped or nonoverlapped and have different dimensions. A subspace can be continuous or discrete (Tokuda et al., 2002). For example, an observation sequence of pitch pattern of speech is composed of one-dimensional continuous values and a discrete symbol which represents "unvoiced."

Suppose the observation space is Ω. State j has N_j subspaces of observations, $\Omega_{j,n} \subseteq \Omega$, $n = 1, \ldots, N_j$, and the probability that its observation is located in the n'th subspace $\Omega_{j,n}$ is $p_{j,n}$, s.t. $\sum_{n=1}^{N_j} p_{j,n} = 1$. Figure 7.2 shows an example, where $\Omega_{i,1}$, $\Omega_{i,2}$, $\Omega_{j,1}$, $\Omega_{j,2}$, and $\Omega_{j,5}$ are two-dimensional continuous subspaces, $\Omega_{i,3}$ and $\Omega_{j,3}$ are one-dimensional continuous subspaces, and $\Omega_{i,4}$ and $\Omega_{j,4}$ are discrete. State i and state j have different division to the observation space Ω.

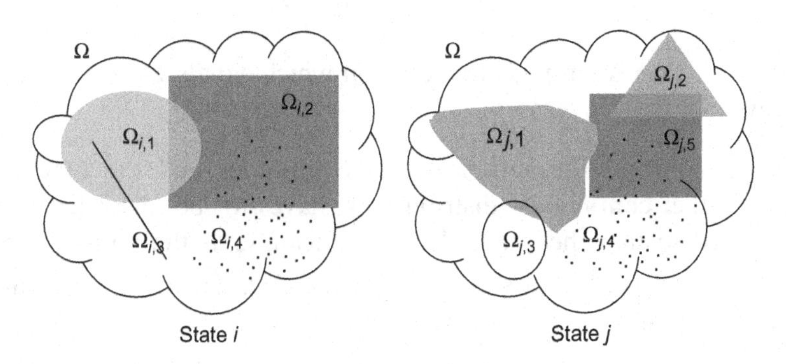

Figure 7.2 Multiple spaces of observations.
The cloud represents the entire observation space Ω. However, the observable area of a state may be only part of the observation space with reduced dimensions. For example, the observable area of state i includes a line of $\Omega_{i,3}$ (one-dimensional), a circle and a square of $\Omega_{i,1}$ and $\Omega_{i,2}$ (two-dimensional), and a set of discrete values of $\Omega_{i,4}$. In contrast, state j has a different observable area of $\Omega_{j,1}$ to $\Omega_{j,5}$.

If subspace $\Omega_{j,n}$ is continuous, then the pdf of continuous observations is denoted by $f_{j,n}(x)$, s.t., $\int_{\Omega_{j,n}} f_{j,n}(x)dx = 1$. If subspace $\Omega_{j,n}$ is discrete, the pmf of discrete observations is denoted by $g_{j,n}(v_k)$, $k = 1, \ldots,$ s.t., $\sum_{v_k \in \Omega_{j,n}} g_{j,n}(v_k) = 1$. Because a subspace may not contain all possible observations of Ω, we define $f_{j,n}(o_t) = 0$ or $g_{j,n}(o_t) = 0$ if $o_t \notin \Omega_{j,n}$. If the subspaces are overlapped, one observation o_t may belong to more than one subspace.

When all subspaces are continuous for given state j, the observation distribution of the state is a mixture of continuous distributions as expressed in Eqn (7.6). Then the model parameters can be estimated by Eqns (7.8) and (7.9). It is important to note that in the case of multispace probability distributions, there may exist $o_t \notin \Omega_{j,n}$ in Eqns (7.8) and (7.9), resulting in $f_{j,n}(o_t) = 0$ and $\gamma_t(j, n) = 0$ based on Eqn (7.12). If $f_{j,n}(x)$ is a Gaussian distribution, the model parameters can be estimated by Eqns (7.10) and (7.11).

When all subspaces are discrete, the mixture $b_j(v_k) = \sum_{n=1}^{N_j} p_{jn} g_{jn}(v_k)$ of observation distributions reduces to a conventional discrete distribution with probabilities $\{b_j(v_k) : k = 1, \ldots, K\}$, which can be estimated by Eqn (3.13). Alternatively, $p_{j,n}$ can be estimated by Eqn (7.8), and $g_{j,n}(v_k)$ can be estimated by

$$\hat{g}_{j,n}(v_k) = \frac{\sum_{t=1}^{T} \gamma_t(j, n) I(o_t = v_k)}{\sum_{t=1}^{T} \gamma_t(j, n)}, \tag{7.13}$$

where $I(o_t = v_k)$ is an indicator function which equals 1 if $o_t = v_k$ and 0 otherwise.

When some of the subspaces are continuous and the others are discrete, the continuous distributions have to be discretized and normalized so that they have the same meaning as the pmfs. Suppose the set of observed values is $\Omega' = \bigcup_{t=1}^{T} \{o_t\} \subset \Omega$. Then the observation probability that $v_k \in \Omega'$ is observed in state j is

$$b_j(v_k) = \sum_{n=1}^{n_j} p_{j,n} f_{j,n}(v_k) \Delta v + \sum_{n=n_j+1}^{N_j} p_{j,n} g_{j,n}(v_k),$$

where n_j is the number of the continuous observation distributions of state j, $N_j - n_j$ is the number of the discrete distributions, and Δv is the interval used for discretizing $f_{j,n}(v)$, for all j and n. To normalize the observation distribution such that $\sum_{v_k \in \Omega'} b_j(v_k) = 1$, we assume

$$\Delta v = \left(1 - \sum_{v_k \in \Omega'} \sum_{n=n_j+1}^{N_j} p_{j,n} g_{j,n}(v_k)\right) \Big/ \sum_{v_k \in \Omega'} \sum_{n=1}^{n_j} p_{j,n} f_{j,n}(v_k), \qquad (7.14)$$

where $\sum_{n=1}^{N_j} p_{j,n} = 1$, and $f_{j,n}(v_k) = 0$ (or $g_{j,n}(v_k) = 0$) if $v_k \notin \Omega_{j,n}$.

7.4 SEGMENTAL MODEL

The HSMMs usually assume that the observation distributions are dependent on the states. These models are not ideal for modeling waveform shapes because only a finite number of piecewise constant shapes, similar to N-gram of symbols, can be modeled. Segmental models relax the constraints by allowing the observation distributions to be dependent on not only the states but also the state durations and their locations within each segment (Krishnamurthy and Moore, 1991; Russell, 1993; Gales and Young, 1993; Katagiri and Lee, 1993; Deng et al., 1994; He and Leich, 1995; Ostendorf et al., 1996; Park et al., 1996; Holmes and Russell, 1999; Yun and Oh, 2000; Ge and Smyth, 2000a; Achan et al., 2005). Dependencies between observations from the same state can thus be described with a parametric trajectory model that changes over time (Russell, 1993; Deng et al., 1994), or a general parametric regression model that allows the mean to be a function of time within each segment (Ge and Smyth, 2000a). A detailed discussion of segmental models can be found in Ostendorf et al. (1996).

In the segmental models, the observation probabilities are defined by $b_{j,d}^{(\tau)}(o_{t+\tau}) \equiv P[o_{t+\tau} | S_{[t+1:t+d]} = j, \tau]$, for $\tau = 1, \ldots, d$, where d denotes the length of the segment or the duration of state j, $o_{t+\tau}$ is the τth observation of the segment $o_{t+1:t+d}$, and $b_{j,d}^{(\tau)}(\cdot)$ the τth distribution region for given state j and the duration d. According to this definition, $b_{j,d_1}^{(\tau_1)}(o_{t+\tau_1})$ and $b_{j,d_2}^{(\tau_2)}(o_{t+\tau_2})$ may be different distributions if $\tau_1 \neq \tau_2$ or $d_1 \neq d_2$. Suppose $h_{j,d}(\tau)$ is a given shape and $\varepsilon_j \sim N(0, \sigma_j^2)$ is a Gaussian noise with zero mean and variance σ^2 for state j and any τ. Then the τth observation within a segment $o_{t+1:t+d}$ of length d given state j is $o_{t+\tau} = h_{j,d}(\tau) + e_j$ with observation distribution

$$b_{j,d}^{(\tau)}(o_{t+\tau}) = \frac{1}{\sqrt{2\pi}\sigma_j}\exp\frac{-(o_{t+\tau} - h_{j,d}(\tau))^2}{2\sigma_j^2}.$$

For example, when $h_{j,d}(\tau) = c_j h_{j,d}(\tau - 1) = c_j^{\tau-1}\mu_j$ with $h_{j,d}(1) = \mu_j$ and $c_j \leq 1$, the observation distribution is

$$b_{j,d}^{(\tau)}(o_{t+\tau}) = \frac{1}{\sqrt{2\pi}\sigma_j}\exp\frac{-(o_{t+\tau} - c_j^{\tau-1}\mu_j)^2}{2\sigma_j^2},$$

where μ_j and σ_j are the parameters of the Gaussian distribution for state j. This model was called "exponentially decay state" by Krishnamurthy and Moore (1991). When $h_{j,d}(\tau) = a_j + c_j\tau$ or $o_{t+\tau} = a_j + c_j\tau + e_j$, the observation is modeled as a linear function of time and

$$b_{j,d}^{(\tau)}(o_{t+\tau}) = \frac{1}{\sqrt{2\pi}\sigma_j}\exp\frac{-(o_{t+\tau} - (a_j + c_j\tau))^2}{2\sigma_j^2},$$

where a_j and c_j are the intercept and slope of a straight line (Kim and Smyth, 2006). Since a waveform can be approximated by several segments of straight lines with different intercepts and slopes, the observations on the waveform can be modeled by a series of states. Each of the states models a segment of straight line plus noise and is allowed only the left-to-right state transition.

Example 7.1

Suppose that the 6′th segment of length $d_6 = 3$ is produced by state 1 with parameters $a_1 = 0$ and $c_1 = 1$, the 7′th segment of length $d_7 = 4$ is by state 2 with $a_2 = 3$ and $c_2 = -0.5$, and the 8′th segment of length $d_8 = 3$ by state 3 with $a_3 = 1$ and $c_3 = 1$. Then the 6′th to 8′th segments containing a sequence of 10 observations represent a waveform like "N" plus noise, as shown in Figure 7.3, where $a_{1,2}$ and $a_{2,3}$ are state transition probabilities, solid curve is the series of observations, and dotted lines are the parametric trajectory $h_{j,d}(\tau) = a_j + c_j\tau$.

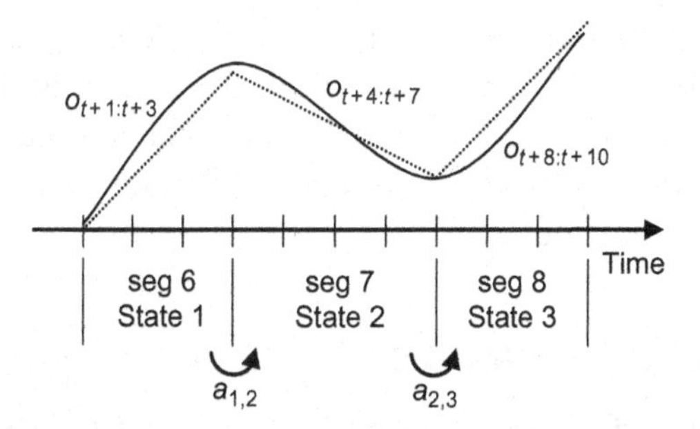

Figure 7.3 Segmental model.
Any observable curve is supposed being produced by a series of states. Each state specifies a straight line that approximates a segment of the curve.

If random effects are added to the segment distributions to model parameter variability across waveforms, then the regression coefficients can be defined by $a_j \sim N(\overline{a}_j, \sigma_j^2)$ and $c_j \sim N(\overline{c}_j, \sigma_j^2)$, and the model can approximate a waveform better (Kim and Smyth, 2006).

Extra-segmental variability associated with a state j can be characterized by a pdf g_j called the *state target PDF* (Russell, 1993; Holmes and Russell, 1995). A target distribution $b_j(\cdot)$ is chosen according to g_j (Austin and Fallside, 1988). Then the joint probability of the segment $o_{t+1:t+d}$ and a particular target $b_j(\cdot)$ given state j is given by (Russell, 1993)

$$P[o_{t+1:t+d}, b_j | S_{[t+1:t+d]} = j] = g_j(b_j) \prod_{\tau=1}^{d} b_{j,d}^{(\tau)}(o_{t+\tau}).$$

Therefore, $P[o_{t+1:t+d} | S_{[t+1:t+d]} = j] = \sum_{b_j} P[o_{t+1:t+d}, b_j | S_{[t+1:t+d]} = j]$.

When the distribution regions $\{b_{j,d}^{(\tau)}(\cdot)\}$ are given, it is a deterministic distribution mapping that associates the τth observation $o_{t+\tau}$, for $\tau = 1, \ldots, d$, in the d-length segment $o_{t+1:t+d}$ with the τth specific region $b_{j,d}^{(\tau)}(\cdot)$. In some other cases, it can be a dynamic distribution mapping that associates the segment $o_{t+1:t+d}$ to a fixed number of regions. Then dynamic programming can be implemented to find the ML mapping. If the observations are assumed conditionally independent given the segment length, a segment model with an unconstrained dynamic mapping is equivalent to a HMM network (Lee et al., 1988).

An extension of the segmental models is defining the observation distributions to be dependent on both the state and its substates (Kwon and Un, 1995). This model is in fact a special case of the original HSMM if we define a complex state that includes both the state and its substate.

7.5 EVENT SEQUENCE MODEL

A hidden semi-Markov event sequence model (HSMESM) was introduced for modeling and analyzing event-based random processes (Thoraval et al., 1994; Thoraval, 2002; Faisan et al., 2002, 2005). At a given time t, an event $o_t = v_k$ can be observed with an occurring probability $1 - e_{S_t}$, or a null observation (missing observation) $o_t = null$ with missing probability e_{S_t}, where S_t is the state at time t. Therefore, the modified observation probabilities in the case that there exist missing observations, denoted by $b_j^+(o_t)$, are given by

$$b_j^+(o_t) = (1 - e_j)b_j(o_t) \cdot I(o_t \in \mathbf{V}) + e_j \cdot I(o_t \notin \mathbf{V}), \quad j \in \mathbf{S},$$

where $\mathbf{V}$ is the set of observable events, and the indicator function $I(x) = 1$ if x is true and 0 otherwise.

This event sequence model is called "state-dependent observation misses" by Yu and Kobayashi (2003b) because the null observation φ is treated as one of the observable events, that is, the full set of observable events is $\mathbf{V} \cup \{\varphi\}$. Yu and Kobayashi (2003b) classified patterns of observation misses into five types. Except the state-dependent observation misses, the other four types are as follows:

1. Output-dependent observation misses: the probability that a given o_t becomes "null" depends on the output value o_t itself. For instance, when the output is too weak (in comparison with noise) at time t, such output may not be observed. In this case, the "output-dependent miss probability" is defined by $e(v_k) = P[\varphi | v_k]$. Then the probability for a null observation $o_t = \varphi$ is $\sum_k b_j(v_k)e(v_k)$.
2. Regular observation misses: the process is sampled at given epochs. Regular or periodic sampling is such a typical example. Some of the outputs of the process may be missed. The missed portion may be significant if the sampling is done infrequently. In this case, the observation probability for a null observation $o_t = \varphi$ is $\sum_k b_j(v_k) = 1$.

3. Random observation misses: the process is sampled at randomly chosen instants. Such observation pattern may apply when the measurement is costly or we are not interested in keeping track of state transitions closely. In this case, some outputs of the process may be missed randomly. If the sampling probability is $1-e$, then the modified observation probability is $b_j^+(o_t)=(1-e)b_j(o_t)I(o_t \in \mathbf{V})+eI(o_t \notin \mathbf{V})$, where e is independent of states.
4. Mismatch between multiple observation sequences: multiple observation sequences are associated with the hidden state sequence, and these observations may not be synchronized to each other. For instance, two sequences $(o_t^{(1)})$ and $(o_t^{(2)})$ are available as the outputs of a HSMM state sequence, but there exits some random delay τ between the two output sequences. Therefore, the observations we can obtain at time t are $o_t^{(1)}$ and $o_{t-\tau}^{(2)}$ though the emissions of a given state at time t are $o_t^{(1)}$ and $o_t^{(2)}$. In this case, delay τ has to be estimated by maximizing the joint likelihood of the two observation sequences.

CHAPTER 8

Variants of HSMMs

This chapter presents variants of HSMMs found in the literature.

8.1 SWITCHING HSMM

A switching HSMM is defined as the concatenation of many HSMMs, with model parameter sets $\lambda_1, \ldots, \lambda_{|Q|}$, each initiated by a different "switching" state $q \in Q$, where the set of states Q defines a Markov chain, as described by Duong et al. (2005a) and Phung et al. (2005a,b). Figure 8.1 shows the switching HSMM, where the sequence of switching states is $q = 1, 4, 2, 3, |Q|$, with state transition probabilities $a_{1,4}$, $a_{4,2}$, $a_{2,3}$, $a_{3,|Q|}$, and in the qth state the process assumes the qth HSMM with the model parameters λ_q. An example of switching HSMM proposed by Sitaram and Sreenivas (1994) is a two-stage inhomogeneous HMM, which is used to capture the variability in speech for phoneme recognition. The first stage models the acoustic and durational variability for all distinct subphonemic segments and the second stage for the whole phoneme.

Assume the entry state of any HSMM is 1, and the departure state of the qth HSMM is M_q. In other words, state M_q cannot transit to a state of the same HSMM but others; and state 1 cannot transit from a state of the same HSMM but others. Therefore, the inter-HSMM transitions are from state M_{q_1} of the q_1th HSMM to state 1 of the q_2th HSMM, for $q_1, q_2 \in Q$. The intra-HSMM transitions are from state i of the qth HSMM to state j of the same HSMM, for $i \neq M_q$ and $j \neq 1$.

Define the forward variable $\alpha_t^{(q)}(j,d)$ as the probability that the partial observation sequence is $o_{1:t}$, the switching state is q, the qth HSMM's state is j, and the state duration is d at time t. Similarly define the backward variable $\beta_t^{(q)}(j,d)$ as the probability that the future observation sequence is $o_{t+1:T}$ for the given qth HSMM's state j and duration d at time t. Then the forward formula (2.6) becomes, for the inter-HSMM transitions,

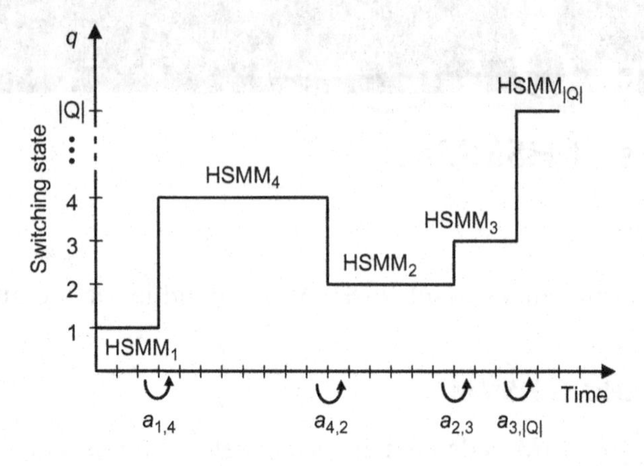

Figure 8.1 Switching HSMM.
The first switching state is 1, and so the set of model parameters that governs this period is $HSMM_1$. In this period, the HSMM state transition probabilities, the state duration distributions, and the observation distributions belong to $HSMM_1$. The length of this period is a random variable governed by the $HSMM_1$ state duration distributions. In this instance, the length of the first period is 3. There may be one to three $HSMM_1$ states occuring in this period. Accordingly, there are three observations. After switching state 1 ends at $t = 3$; it transits to switching state 4 with switching state transition probability $a_{1,4}$. Then it enters switching state 4 at $t = 4$ and starts being governed by the set of model parameters $HSMM_4$. The length of the second period is again a random variable. In this instance, it is 8. Therefore, there may be one to eight $HSMM_4$ states present in the second period with exactly eight observations.

$$\alpha_t^{(q)}(1,d) = \sum_l \left(\sum_h \alpha_{t-d}^{(l)}(M_l,h) \right) \cdot a_{l,q} \cdot b_{1,d}^{(q)}(o_{t-d+1:t}), \tag{8.1}$$

and for the intra-HSMM transitions,

$$\alpha_t^{(q)}(j,d) = \sum_{i \neq j \text{ or } M_q} \sum_h \alpha_{t-d}^{(q)}(i,h) \cdot a_{(i,h)(j,d)}^{(q)} \cdot b_{j,d}^{(q)}(o_{t-d+1:t}), \quad j \neq 1, \tag{8.2}$$

for all $q \in Q$, $j \in \mathbf{S}$, $d \in \mathbf{D}$, and $t = 1, \ldots, T$, where $a_{l,q}$ is the state transition probability from switching state l to switching state q, $a_{(i,h)(j,d)}^{(q)}$ is the state transition probability from state i having duration h to state j having duration d within the qth HSMM, and $b_{j,d}^{(q)}(o_{t-d+1:t})$ are the observation probabilities of the qth HSMM. Similarly, the backward formula (2.7) becomes

$$\beta_t^{(q)}(j,d) = \sum_{i \neq j \text{ or } 1} \sum_h a_{(j,d)(i,h)}^{(q)} \cdot b_{i,h}^{(q)}(o_{t+1:t+h}) \cdot \beta_{t+h}^{(q)}(i,h), \quad j \neq M_q,$$

$$\beta_t^{(q)}(M_q,d) = \sum_l \sum_h a_{q,l} \cdot b_{1,h}^{(l)}(o_{t+1:t+h}) \cdot \beta_{t+h}^{(l)}(1,h). \tag{8.3}$$

The smoothed probabilities can be calculated by, for instance,

$$\eta_t^{(q)}(j,d) = \alpha_t^{(q)}(j,d)\beta_t^{(q)}(j,d), \tag{8.4}$$

$$\gamma_t^{(q)}(j) = \sum_{\tau \geq t} \sum_{d=\tau-t+1}^{D} \eta_\tau^{(q)}(j,d), \tag{8.5}$$

similar to Eqns (2.10)–(2.13).

8.2 ADAPTIVE FACTOR HSMM

Adaptive factor HSMM is assumed to have variable model parameters for its parametric distributions of observations and/or state durations. For example, the mean μ_i of $b_i(v_k)$ given state i is changed for different speakers, such that for the f'th speaker the mean becomes $a_i^{(f)} + c_i^{(f)}\mu_i$, $f = 1,\ldots,F$, where $\{a_i^{(f)}, c_i^{(f)}\}$ are adaptive factors, μ_i the common parameters for all speakers, and F the total number of speakers.

Let $\mathbf{o}^{(f)} = (o_1^{(f)}, o_2^{(f)},\ldots, o_{T_f}^{(f)})$ be the f'th observation sequence of length T_f, for $f = 1,\ldots,F$, λ be the set of the common parameters of the HSMM, and $\Lambda = \{\Lambda^{(1)},\ldots,\Lambda^{(F)}\}$ be the set of the adaptive factors. For example, in the last example, $\mu_i \subset \lambda$ and $(a_i^{(f)}, c_i^{(f)}) \subset \Lambda^{(f)}$. Then the model parameters can be jointly estimated by

$$\{\hat{\lambda}, \hat{\Lambda}\} = \arg\max_{\lambda,\Lambda} P[\mathbf{o}^{(1)},\ldots,\mathbf{o}^{(F)}|\lambda,\Lambda] = \arg\max_{\lambda,\Lambda} \prod_{f=1}^{F} P[\mathbf{o}^{(f)}|\lambda,\Lambda^{(f)}]. \tag{8.6}$$

Following a similar procedure in deriving Algorithm 3.2 for multiple observation sequences, we can get the estimation formulas for the model parameters of the adaptive factor HSMM. That is, similar to Eqns (3.10), (7.1) or (7.2), the estimation formula for the parameters of the observation distributions is

$$\max_{\lambda,\Lambda} \sum_{f=1}^{F} \sum_j \sum_t \overline{\gamma}_t^{(f)}(j) \log b_j^{(f)}(o_t^{(f)}; \lambda, \Lambda^{(f)}),$$

where $\overline{\gamma}_t^{(f)}(j) = P[S_t = j|\mathbf{o}^{(f)}, \lambda, \Lambda^{(f)}]$ are the smoothed probabilities calculated from the fth observation sequence $\mathbf{o}^{(f)}$ given the model parameters $\{\lambda, \Lambda^{(f)}\}$.

Specific examples for the joint estimation of the model parameters are given by Yamagishi and Kobayashi (2005) and Yamazaki et al. (2005), where $o_t^{(f)}$, $a_i^{(f)}$, and μ_i are vectors, and $c_i^{(f)}$ is a matrix.

8.3 CONTEXT-DEPENDENT HSMM

In different contexts a state can have different observation distributions and duration distributions. The model can be described by a series of concatenated HSMMs, each of which may have different model parameters. Different from the switching HSMM, the vector of contextual factors that determines the choice of the model parameters has infinite support. We will call a vector of contextual factors as a contextual vector in the rest of this section.

For example, a linguistic specification derived from a text includes phonetic and prosodic properties of the current, preceding, and following segments, which can be considered a vector of contextual factors. Acoustic features (e.g., spectrum, excitation, and duration) characterizing the speech waveform of the text are dependent on a series of contextual vectors. In other words, the same acoustic unit (e.g., a word or a phone) may have different acoustic features in different contexts. Each acoustic unit can be characterized by a context-dependent HSMM, and the acoustic features of the text can be described by a series of concatenated HSMMs.

It is always assumed that the series of contextual vectors assigned to an observation sequence is given, though the states are still hidden/unobservable. For example, the text for a speech is always known in the area of speech synthesis, and so the series of contextual vectors corresponding to a series of acoustic features (i.e., an observation sequence) is known.

Figure 8.2 shows the context-dependent HSMM, where the observation sequence and the series of contextual vectors, $\mathbf{f}_{1:5}$, are given. The series of contextual vectors is corresponding to the observation sequence, but is not aligned to the observation sequence as well as the state sequence. The state sequence and the state durations are hidden and must be estimated based on the given observation sequence and the given series of contextual vectors. In the nth HSMM, the set of model parameters is chosen according to the nth contextual vector, $\mathbf{f}_n$. This model is similar to a switching HSMM if the series of contextual vectors is considered to be a given chain of switching states. The difference from the switching HSMM is that $\mathbf{f}_n$ is given and not finite and countable.

Therefore, a contextual vector must be discretized so that the closed contextual vectors are classified into the same class, and each class is

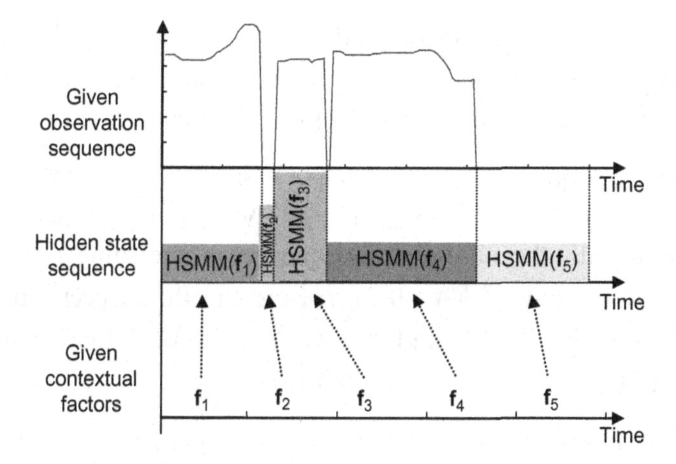

Figure 8.2 Context-dependent HSMM.
The series of contextual vectors, (f_1, ..., f_5) is given. The corresponding observation sequence $o_{1:T}$ is given. However, the underlying state sequence is hidden and the state jump times or durations are unknown. Because the series of contextual vectors is given, the sets of model parameters are known to be in the order HSMM(f_1),..., HSMM(f_5).

treated as a switching state. Suppose the contextual vector $\mathbf{f}_n$ is classified as class q_n. The series of classes $q_{1:N}$ corresponding to the given series of contextual vectors $\mathbf{f}_{1:N}$ is then treated as the given chain of switching states, where N is the length of the series. The transition probability a_{q_{n-1},q_n} from q_n to q_{n+1} is always 1 because $q_{1:N}$ is given. The forward–backward formulas (8.1)–(8.3) specified for the switching HSMM become

$$\alpha_t^{(q_n)}(j,d) = \sum_{i \neq j \text{ or } M_{q_n}} \sum_h \alpha_{t-d}^{(q_n)}(i,h) \cdot a_{(i,h)(j,d)}^{(q_n)} \cdot b_{j,d}^{(q_n)}(o_{t-d+1:t}), \quad j \neq 1,$$
$$\alpha_t^{(q_n)}(1,d) = \left(\sum_h \alpha_{t-d}^{(q_{n-1})}(M_{q_{n-1}},h) \right) \cdot b_{1,d}^{(q_n)}(o_{t-d+1:t}), \tag{8.7}$$

and

$$\beta_t^{(q_n)}(j,d) = \sum_{i \neq j \text{ or } 1} \sum_h a_{(j,d)(i,h)}^{(q_n)} \cdot b_{i,h}^{(q_n)}(o_{t+1:t+h}) \cdot \beta_{t+h}^{(q_n)}(i,h), \quad j \neq M_{q_n}$$
$$\beta_t^{(q_n)}(M_{q_n},d) = \sum_h b_{1,h}^{(q_{n+1})}(o_{t+1:t+h}) \cdot \beta_{t+h}^{(q_{n+1})}(1,h). \tag{8.8}$$

In considering that each contextual vector may be assigned for a left-o-right HSMM_{q_n}, which produces at least M_{q_n} observations, the constraints for Eqns (8.7) and (8.8) are $t \geq \sum_{m=1}^{n-1} M_{q_m}$ and $T - t \geq \sum_{m=n+1}^{N} M_{q_m}$. For example, when a five-state left-to-right HSMM with no skip topology is applied as the model of a phoneme, $M_{q_n} = 5$.

Using Eqn (8.5), the smoothed probabilities, $\overline{\gamma}_t^{(q_n)}(j) = \gamma_t^{(q_n)}(j)/P[o_{1:T}|\lambda]$, are calculated. We note that because $q_{1:N}$ is not aligned with the observation sequence, at the same t there may exist $\overline{\gamma}_t^{(q_n)}(j) > 0$ and $\overline{\gamma}_t^{(q_m)}(j) > 0$ for $n \neq m$. In other aspect, due to the constraints $t \geq \sum_{m=1}^{n-1} M_{q_m}$ and $T - t \geq \sum_{m=n+1}^{N} M_{q_m}$, there may exist $\overline{\gamma}_t^{(q_2)}(j) = 0$. For example, $\overline{\gamma}_t^{(q_2)}(j) = 0$, for $t \leq M_{q1}$.

Now we discuss how to estimate the model parameters of the observation distributions for a class of contextual vectors. The model parameters for the state duration distributions for a class of contextual vectors can be similarly estimated.

From Eqn (7.2) we can see that the model parameters for the observation distribution of state j of class l can be estimated by maximizing the log-likelihood function

$$\max_{\{\varphi_j^{(l)}\}} \sum_{t,n} \overline{\gamma}_t^{(q_n)}(j) I(q_n = l) \log b_j(o_t, \varphi_j^{(l)}), \quad l = 1, \ldots, L, \; j \in \mathbf{S}, \tag{8.9}$$

where $I(x)$ is the indicator function, $\varphi_j^{(l)}$ is the set of model parameters to be estimated for the observation distribution $b_j(o_t, \varphi_j^{(l)})$ of state j with the contextual vectors belonging to class l, and L is the number of classes. Though Eqn (8.9) sums over from time 1 to T, it does not actually include all the times because some $\overline{\gamma}_t^{(l)}(j)$ may be 0s.

Suppose $b_j(o_t, \varphi_j^{(l)})$ belong to the exponential family, that is,

$$b_j(o_t, \varphi_j^{(l)}) = \frac{1}{B(\varphi_j^{(l)})} \xi(o_t) \exp\left(-\sum_{p=1}^{P} \varphi_{j,p}^{(l)} S_p(o_t)\right), \tag{8.10}$$

where P is the number of natural parameters, $\varphi_{j,p}^{(l)}$ is the p'th natural parameter for state j of class l, $\varphi_j^{(l)} = (\varphi_{j,1}^{(l)}, \ldots, \varphi_{j,P}^{(l)})$, $S_p(o_t)$ and $\xi(o_t)$ are sufficient statistic, and $B(\varphi_j^{(l)})$ is a normalizing term satisfying

$$B(\varphi_j^{(l)}) = \int_0^{\infty} \xi(x) \exp\left(-\sum_{p=1}^{P} \varphi_{j,p}^{(l)} S_p(x)\right) dx. \tag{8.11}$$

Then the maximization problem of Eqn (8.9) becomes

$$\max_{\{\varphi_j^{(l)}\}} \sum_{t,n} \overline{\gamma}_t^{(q_n)}(j) I(q_n = l) \left(-\log B(\varphi_j^{(l)}) - \sum_{p=1}^{P} \varphi_{j,p}^{(l)} S_p(o_t) + \log \xi(o_t) \right). \tag{8.12}$$

Since the exponential family is log-concave, the global maximum can be found by setting the derivative equal to zero, yielding the maximum likelihood equations, that is,

$$\sum_{t,n} \overline{\gamma}_t^{(q_n)}(j) I(q_n = l) \left(\frac{\partial \log B(\varphi_j^{(l)})}{\partial \varphi_{j,p}^{(l)}} + S_p(o_t) \right) = 0,$$

where

$$\begin{aligned} -\frac{\partial \log B(\varphi_j^{(l)})}{\partial \varphi_{j,p}^{(l)}} &= \frac{1}{B(\varphi_j^{(l)})} \int_0^{\infty} \xi(x) S_p(x) \exp\left(-\sum_{p=1}^{P} \varphi_{j,p}^{(l)} S_p(x) \right) dx \\ &= \int_0^{\infty} b_j(x, \varphi_j^{(l)}) S_p(x) dx \\ &= E(S_p(x) | \varphi_j^{(l)}) \end{aligned}$$

is the expected value taken with respect to the exponential family member. Denote $E_p(\varphi_j^{(l)}) = E(S_p(x)|\varphi_j^{(l)})$. Then, the new parameters $\hat{\varphi}_j^{(l)} = (\hat{\varphi}_{j,1}^{(l)}, \ldots, \hat{\varphi}_{j,P}^{(l)})$ as well as the expected values $E_1(\hat{\varphi}_j^{(l)}), \ldots, E_P(\hat{\varphi}_j^{(l)})$ can be estimated by solving the following equations:

$$E_p(\hat{\varphi}_j^{(l)}) = \frac{\sum_{t,n} \overline{\gamma}_t^{(q_n)}(j) I(q_n = l) S_p(o_t)}{\sum_{t,n} \overline{\gamma}_t^{(q_n)}(j) I(q_n = l)}, \tag{8.13}$$

for $p = 1, \ldots, P$, $l = 1, \ldots, L$, and $j \in \mathbf{S}$. Substituting the optimal solution $\hat{\varphi}_j^{(l)}$ and $E_p(\hat{\varphi}_j^{(l)})$ in Eqn (8.12) yields the maximum log-likelihood

$$\sum_{t,n} \overline{\gamma}_t^{(q_n)}(j) I(q_n = l) \left(-\log B(\hat{\varphi}_j^{(l)}) - \sum_{p=1}^{P} \hat{\varphi}_{j,p}^{(l)} E_p(\hat{\varphi}_j^{(l)}) + \log \xi(o_t) \right) \tag{8.14}$$

or

$$\sum_{t,n} \overline{\gamma}_t^{(q_n)}(j) I(q_n = l) \log \left[\frac{1}{B(\hat{\varphi}_j^{(l)})} \xi(o_t) \exp\left(-\sum_{p=1}^{P} \hat{\varphi}_{j,p}^{(l)} E_p(\hat{\varphi}_j^{(l)}) \right) \right]. \tag{8.15}$$

Suppose the contextual vectors of class l can be divided with finer grain into K subclasses $l_1, \ldots, l_K$. In the given series of classes $q_{1:N}$, some $q_n = l$ become $q_n = l_k$ dependent on the contextual vector $\mathbf{f}_n$, $k \in \{1,\ldots,K\}$. Denote this new series of classes by $q'_{1:N}$. If the given model parameters for the subclasses $l_1,\ldots, l_K$ are the same as class l, then the smoothed probabilities $\sum_{k=1}^{K}\sum_{t,n}\overline{\gamma}_t^{(q_n)}(j)I(q'_n = l_k) = \sum_{t,n}\overline{\gamma}_t^{(q_n)}(j)I(q_n = l)$, for $j \in \mathbf{S}$, where $\overline{\gamma}_t^{(q_n)}$ are determined by $q_{1:N}$ instead of $q'_{1:N}$. Therefore, the maximum log-likelihood for subclass l_k is

$$\sum_{t,n}\overline{\gamma}_t^{(q_n)}(j)I(q'_n = l_k)\left(-\log B(\hat{\varphi}_j^{(l_k)}) - \sum_{p=1}^{P}\hat{\varphi}_{j,p}^{(l_k)}E_p(\hat{\varphi}_j^{(l_k)}) + \log \xi(o_t)\right).$$

By substituting $\sum_{k=1}^{K}\sum_{t,n}\overline{\gamma}_t^{(q_n)}(j)I(q'_n = l_k) = \sum_{t,n}\overline{\gamma}_t^{(q_n)}(j)I(q_n = l)$ in Eqn (8.14), the increment of the log-likelihood function due to the class division can be determined by

$$\sum_{k=1}^{K}\left[\log\frac{B(\hat{\varphi}_j^{(l)})}{B(\hat{\varphi}_j^{(l_k)})} + \sum_{p=1}^{P}(\hat{\varphi}_{j,p}^{(l)}E_p(\hat{\varphi}_j^{(l)}) - \hat{\varphi}_{j,p}^{(l_k)}E_p(\hat{\varphi}_j^{(l_k)}))\right]\sum_{t,n}\overline{\gamma}_t^{(q_n)}(j)I(q'_n = l_k). \tag{8.16}$$

Example 8.1 Observation with Gaussian Distributions

The Gaussian distribution $\frac{1}{\sqrt{2\pi}\sigma}\exp\left(-\frac{(x-\mu)^2}{2\sigma^2}\right) = \frac{1}{\sqrt{2\pi}\sigma}\exp\left(-\frac{\mu^2}{2\sigma^2}\right)\cdot \exp\left(-\frac{x^2-2x\mu}{2\sigma^2}\right)$ can be expressed in the form of Eqn (8.10) with parameters as follows: $P = 2$, $S_1(x) = x$, $S_2(x) = x^2$, $\varphi_{j,1}^{(l)} = \frac{-\mu}{\sigma^2}$, $\varphi_{j,2}^{(l)} = \frac{1}{2\sigma^2}$, $\xi(x) = 1$, $B(\varphi_j^{(l)}) = \sqrt{2\pi}\sigma\exp\left(\frac{\mu^2}{2\sigma^2}\right)$, $E_1(\varphi_j^{(l)}) = \mu$, and $E_2(\varphi_j^{(l)}) = \sigma^2 + \mu^2$. We have $\sum_{p=1}^{2}\varphi_{j,p}^{(l)}E_p(\varphi_j^{(l)}) = -\frac{\mu^2}{2\sigma^2} + \frac{1}{2}$ and $\log B(\varphi_j^{(l)}) + \sum_{p=1}^{2}\varphi_{j,p}^{(l)}E_p(\varphi_j^{(l)}) = \log\sigma + \frac{1}{2}\log(2\pi) + \frac{1}{2}$. Then the increment of the likelihood function by clustering the class l into K nonoverlapped subclasses l_1 to l_K is

$$\sum_{k=1}^{K}\log\frac{\hat{\sigma}_l}{\hat{\sigma}_{l_k}}\sum_{t,n}\overline{\gamma}_t^{(q_n)}(j)I(q'_n = l_k), \tag{8.17}$$

where $\hat{\mu}_{l_k} = \frac{\sum_{t,n}\overline{\gamma}_t^{(q_n)}(j)I(q'_n = l_k)o_t}{\sum_{t,n}\overline{\gamma}_t^{(q_n)}(j)I(q'_n = l_k)}$ and $\hat{\sigma}_{l_k}^2 = \frac{\sum_{t,n}\overline{\gamma}_t^{(q_n)}(j)I(q'_n = l_k)o_t^2}{\sum_{t,n}\overline{\gamma}_t^{(q_n)}(j)I(q'_n = l_k)} - \hat{\mu}_{l_k}^2$, for $k = 1, \ldots, K$. Similar formulas for HMMs can be found in Odell (1995) and Khorram et al. (2015).

Because one can obtain only a limited number of observation sequences, they are extremely sparse compared with the huge space of contextual factors. A decision tree used to classify the contextual vectors into a few of classes is critical, so that only a few sets of model parameters are required to be estimated from the sparse samples. Since the decision tree is unknown, we have to estimate the decision tree along with the sets of model parameters.

Use $D_l(\mathbf{f}_n) = 1$ to indicate that the decision tree classifies the contextual vector $\mathbf{f}_n$ into leaf l; otherwise it is 0, $n = 1, \ldots, N$. Equation (8.16) can be equivalently expressed as

$$\sum_{k=1}^{K}\left[\log\frac{B(\hat{\boldsymbol{\varphi}}_j^{(l)})}{B(\hat{\boldsymbol{\varphi}}_j^{(l_k)})} + \sum_{p=1}^{P}(\hat{\varphi}_{j,p}^{(l)}E_p(\hat{\boldsymbol{\varphi}}_j^{(l)}) - \hat{\varphi}_{j,p}^{(l_k)}E_p(\hat{\boldsymbol{\varphi}}_j^{(l_k)}))\right]\sum_{t,n}\overline{\gamma}_t^{(q_n)}(j)D_{l_k}(\mathbf{f}_n). \tag{8.18}$$

Without losing generality, we discuss how to build a binary tree. First, let the decision tree have only one leaf (root), that is, $l = 0$. In this case, $D_0(\mathbf{f}_n) = 1$, for all n. In other words, different contextual vectors use the same set of model parameters. This is the original HSMMs regardless the contextual factors. Use the re-estimation algorithm of Algorithm 3.1 to estimate the set of model parameters until the likelihood function reaches the maximum. At this time, the set of model parameters cannot be further improved because the maximum likelihood has been reached. Then calculate the smoothed probabilities $\overline{\gamma}_t^{(q_n)}(j)$ over all the observation sequences, where $q_n = 0$, for all n. Now the class $l = 0$ is to be divided into two subclasses. Use Eqn (8.18) to calculate the increment of the log-likelihood function for $K = 2$. Select the best $D_1(\mathbf{f}_n)$ and $D_2(\mathbf{f}_n)$ so that the increment of the log-likelihood function is maximized. Then the leaf $l = 0$ is split into two leaves $l_1 = 1$ and $l_2 = 2$ according to $D_1(\mathbf{f}_n)$ and $D_2(\mathbf{f}_n)$, and the set $\mathbf{C}_0 = \bigcup_{n=1}^{N}\{\mathbf{f}_n\}$ of the contextual vectors is divided into $\mathbf{C}_1$ and $\mathbf{C}_2$ with $\mathbf{C}_1 \cup \mathbf{C}_2 = \mathbf{C}_0$. That is, $D_k(\mathbf{f}_n) = 1$, if $\mathbf{f}_n \in \mathbf{C}_k$; otherwise, $D_k(\mathbf{f}_n) = 0$, where $k = 1,2$.

Now the decision tree has two leafs that cluster the contextual vectors into two classes. This means there are two sets of model parameters corresponding to two HSMMs, which are concatenated to form the process. Use the forward–backward formulas (8.7) and (8.8) and the re-estimation algorithm of Algorithm 3.1 to update the model parameters until the likelihood function reaches the maximum. Then calculate the smoothed probabilities $\overline{\gamma}_t^{(q_n)}(j)$ based on Eqn (8.5), where $q_n = 1$ or 2. Try to split $l = 1$ and $l = 2$, respectively, and use Eqn (8.18) to find the maximum increment of the log-likelihood function for $K = 2$. Suppose the best choice is $D_{l_1}(\mathbf{f}_n)$ and $D_{l_2}(\mathbf{f}_n)$, for $\mathbf{f}_n \in \mathbf{C}_l$, which maximizes the increment of the log-likelihood function. After that, leaf l is split into two new leaves l_1 and l_2. The total number of leaves becomes three. That is, there are three sets of model parameters corresponding to three HSMMs. This procedure, that is, "split-update" of model parameters, is repeated until the maximum log-likelihood is reached or the contextual vectors cannot be split with finer grain.

A simple, but approximated, approach to build the decision tree is assuming that the smoothed probabilities $\{\overline{\gamma}_t^{(q_n)}(j)\}$ are fixed during the clustering procedure (Odell, 1995; Khorram et al., 2015). At first, Algorithm 3.1 is implemented until the maximum likelihood is reached. Then the smoothed probabilities $\{\overline{\gamma}_t^{(q_n)}(j)\}$ are calculated and assumed to be fixed in the later clustering procedure, where $q_n = 0$, for all n. During the clustering procedure, Eqn (8.18) is repeatedly applied to split the decision tree until a termination criterion is satisfied. Suppose the resulting tree has L leaves. Then, this procedure is equivalent to applying Eqn (8.18) once with $K = L$.

If there are multiple observation sequences we can compute the smoothed probabilities for each of the observation sequences. Denote $\overline{\gamma}_t^{(q_n^{(r)})}(j)$ as the smoothed probabilities computed for the given model parameters λ and the given rth observation sequence $o_{1:T_r}^{(r)}$ of length T_r, where the model parameters λ are the same for all the observation sequences, but the given contextual vector sequences $\mathbf{f}_{1:N_r}^{(r)}$ of length N_r are different. Then the new parameters $\hat{\varphi}_j^{(l)} = (\hat{\varphi}_{j,1}^{(l)}, \ldots, \hat{\varphi}_{j,P}^{(l)})$ as well as the expected values $E_1(\hat{\varphi}_j^{(l)}), \ldots, E_P(\hat{\varphi}_j^{(l)})$ can be estimated by solving the following equations similar to (8.13):

$$E_p(\hat{\varphi}_j^{(l)}) = \frac{\sum_r \sum_{n=1}^{N_r} \sum_{t=1}^{T_r} \overline{\gamma}_t^{(q_n^{(r)})}(j) I(q_n^{(r)} = l) S_p(o_t)}{\sum_r \sum_{n=1}^{N_r} \sum_{t=1}^{T_r} \overline{\gamma}_t^{(q_n^{(r)})}(j) I(q_n^{(r)} = l)},$$

and similar to Eqn (8.18), the increment of the log-likelihood function due to the class division can be determined by

$$\sum_{k=1}^{K}\left[\log\frac{B(\hat{\varphi}_j^{(l)})}{B(\hat{\varphi}_j^{(l_k)})}+\sum_{p=1}^{P}(\hat{\varphi}_{j,p}^{(l)}E_p(\hat{\varphi}_j^{(l)})-\hat{\varphi}_{j,p}^{(l_k)}E_p(\hat{\varphi}_j^{(l_k)}))\right]$$

$$\sum_{r}\sum_{n=1}^{N_r}\sum_{t=1}^{T_r}\overline{\gamma}_t^{(q_n^{(r)})}(j)D_{l_k}(\mathbf{f}_n^{(r)}),$$

where the class label $q_n^{(r)}$ is corresponding to the contextual vector $\mathbf{f}_n^{(r)}$.

8.4 MULTICHANNEL HSMM

Multichannel HSMM was proposed to model multiple interacting processes (Natarajan and Nevatia, 2007b). In contrast to the basic HSMM that a process has a single state at any instant, this extension generalizes the HSMM state to be a vector $\mathbf{S}_t = (S_t^{(1)}, \ldots, S_t^{(C)})$ representing the states of multiple processes. Each of the states, $S_t^{(c)}$, is dependent on all previous states $\mathbf{S}_{t-1} = (S_{t-1}^{(1)}, \ldots, S_{t-1}^{(C)})$. Corresponding to the observation sequence, $\mathbf{O}^{(c)}$, of the c'th process is a hidden semi-Markov state sequence, $S_{[1:d_1]}^{(c)}, \ldots, S_{[T-d_n+1:T]}^{(c)}$, given a set of model parameters, $\lambda^{(c)}$, where T is the length of the observation sequence. Each observation $o_t^{(c)}$ is generally dependent on all current states $\mathbf{S}_t = (S_t^{(1)}, \ldots, S_t^{(C)})$. Figure 8.3 shows the multichannel HSMM, where $\mathbf{s}_t = (s_t^{(1)}, \ldots, s_t^{(C)})$, $\mathbf{d}_t = (d_t^{(1)}, \ldots, d_t^{(C)})$, $\mathbf{o}_t = (o_t^{(1)}, \ldots, o_t^{(C)})$, $s_t^{(c)}$ is the state of channel c at time t, $d_t^{(c)}$ is the time having been spent in state $s_t^{(c)}$ by time t, and $o_t^{(c)}$ is the observation of channel c at time t.

Though this model can be realized in the framework of inhomogeneous HMM, the state space will be $\prod_{c=1}^{C} M_c$, where M_c is the number of states $S_t^{(c)}$ takes. To reduce the complexity of the model, some simplifying assumptions should be made for the transition probabilities. Assume the state transition probabilities are $P[\mathbf{s}_{t+1}, \mathbf{d}_{t+1}|\mathbf{s}_t, \mathbf{d}_t, \lambda]$, and the observation probabilities are $P[\mathbf{o}_t|\mathbf{s}_t, \lambda]$. Then, each channel is usually allowed to evolve independently, that is,

$$P[\mathbf{s}_{t+1}, \mathbf{d}_{t+1}|\mathbf{s}_t, \mathbf{d}_t, \lambda] = \prod_{c=1}^{C} P[s_{t+1}^{(c)}, d_{t+1}^{(c)}|\mathbf{s}_t, \mathbf{d}_t, \lambda]$$

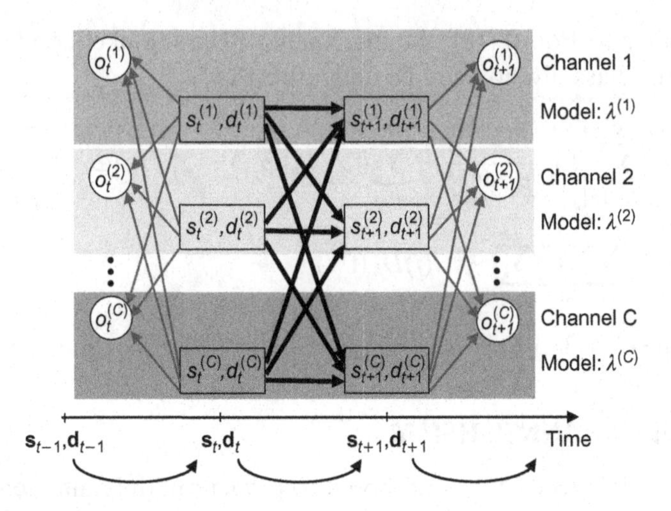

Figure 8.3 Multichannel HSMM.
There are C *channels. In each channel, there is an observation sequence and a hidden semi-Markov state sequence. Different channels have different observation and state sequences for different given model parameters. However, each observation and each state in one channel are dependent on all the states of the channels. For example, the state* $s_{t+1}^{(1)}$ *of channel* 1 *is dependent on all the previous states* $s_t^{(1)}, \ldots, s_t^{(C)}$*, and the observation* $o_{t+1}^{(1)}$ *is dependent on all the current states* $s_{t+1}^{(1)}, \ldots, s_{t+1}^{(C)}$*. The channels are not synchronized because the states* $s_{t+1}^{(1)}, \ldots, s_{t+1}^{(C)}$ *may have different durations, starting/ending at different times.*

and

$$P[\mathbf{o}_t|\mathbf{s}_t, \lambda] = \prod_{c=1}^{C} P[o_t^{(c)}|s_t^{(c)}, \lambda].$$

To further simplify the expressions, $P[s_{t+1}^{(c)}, d_{t+1}^{(c)}|\mathbf{s}_t, \mathbf{d}_t, \lambda]$ is approximated by (Natarajan and Nevatia, 2007b)

$$P[s_{t+1}^{(c)}, d_{t+1}^{(c)}|\mathbf{s}_t, \mathbf{d}_t, \lambda] \sim \prod_{c'=1}^{C} P[s_{t+1}^{(c)}, d_{t+1}^{(c)}|s_t^{(c')}, d_t^{(c')}, \lambda]$$

or

$$P[s_{t+1}^{(c)}, d_{t+1}^{(c)}|\mathbf{s}_t, \mathbf{d}_t, \lambda] = \sum_{c'=1}^{C} w_{c,c'} P[s_{t+1}^{(c)}, d_{t+1}^{(c)}|s_t^{(c')}, d_t^{(c')}, \lambda],$$

where $w_{c,c'}$ are the weights.

According to this simplified assumption, the current state–duration pair $(s_{t+1}^{(c)}, d_{t+1}^{(c)})$ is dependent on the previous state–duration pair $(s_t^{(c')}, d_t^{(c')})$. There are MD current state–duration pairs for given c and MD previous state–duration pairs for given c'. Therefore, the computational complexity for evaluating the forward or backward variables at each time step is $O(M^2D^2C^2)$.

8.5 SIGNAL MODEL OF HSMM

For a discrete HSMM, the set of states is usually assumed as $\mathbf{S} = \{1,\ldots, M\}$ or $\mathbf{S} = \{s_1,\ldots, s_M\}$, and the set of observable values/symbols as $\mathbf{V} = \{v_1,\ldots, v_K\}$. The state of the process at time t is denoted by $S_t \in \mathbf{S}$, and the corresponding observation by $O_t \in \mathbf{V}$. Because the specific values of s_j and v_k do not affect the implementation of the forward–backward algorithms of HSMMs, one usually uses their indices to retrieve the model parameters such as a_{ij} and $b_j(k)$. Therefore, we can equivalently let the state of the process at time t be a $M \times 1$ vector $\mathbf{X}_t$ which takes its values from the set $\{\mathbf{e}_1,\ldots,\mathbf{e}_M\}$, and the corresponding observation be a $K \times 1$ vector $\mathbf{Y}_t$ taking its values from the set $\{\mathbf{v}_1,\ldots,\mathbf{v}_K\}$, where $\mathbf{e}_j$ is a $M \times 1$ vector with unity as the j'th element and zeros elsewhere, and $\mathbf{v}_k$ is a $K \times 1$ vector with unity as the k'th element and zeros elsewhere. The advantage of these expressions is that one can use expectations to compute various (filtering, predicting, and smoothing) probabilities (Elliott et al., 1995, 2013). For example,

$$E\left[\mathbf{X}_t^T \mathbf{e}_j\right] = \sum_i \mathbf{e}_i^T \mathbf{e}_j P[\mathbf{X}_t = \mathbf{e}_i] = P[\mathbf{X}_t = \mathbf{e}_j],$$

$$E[\mathbf{X}_{t+1}|\mathbf{X}_t] = \sum_j \mathbf{e}_j P[\mathbf{X}_{t+1} = \mathbf{e}_j|\mathbf{X}_t] = \mathbf{A}^T\mathbf{X}_t,$$

$$E[\mathbf{Y}_t|\mathbf{X}_t] = \sum_k \mathbf{v}_k P[\mathbf{Y}_t = \mathbf{v}_k|\mathbf{X}_t] = \mathbf{B}^T\mathbf{X}_t,$$

where the superscript "T" represents transpose of a matrix or vector, $\mathbf{A}$ is the state transition probability matrix with $a_{ij} = P[\mathbf{X}_{t+1} = \mathbf{e}_j|\mathbf{X}_t = \mathbf{e}_i]$, and $\mathbf{B}$ is the observation probability matrix with $b_{ik} = P[\mathbf{Y}_t = \mathbf{v}_k|\mathbf{X}_t = \mathbf{e}_i]$. This model was first introduced by Elliott et al. (1995) as a signal model of HMM and then extended to HSMM by Azimi et al. (2003, 2004, 2005) and Elliott et al. (2013).

In the signal model of HSMM, the hidden state $(\mathbf{X}_t = \mathbf{e}_i, d_t = d)$ can transit to either $(\mathbf{X}_{t+1} = \mathbf{e}_i, d_{t+1} = d_t + 1)$ or $(\mathbf{X}_{t+1} = \mathbf{e}_j, d_{t+1} = 1)$ for $i \neq j$, that is,

$$\mathbf{d}_{t+1} = Diag(\mathbf{X}_{t+1})(\mathbf{d}_t + \mathbf{1}),$$

where $\mathbf{d}_t$ is the duration vector of size $M \times 1$ with all elements equal zeros except one element (e.g., the ith element) equals $d_t \geq 1$, denoting the duration spent in state $\mathbf{X}_t = \mathbf{e}_i$ prior to time t. $Diag(\mathbf{X}_{t+1})$ changes the vector $\mathbf{X}_{t+1}$ into a diagonal matrix; $\mathbf{1}$ is a $M \times 1$ vector with all unity elements. The state $\mathbf{X}_t$ then transits to the next state $\mathbf{X}_{t+1}$ satisfying

$$\mathbf{X}_{t+1} = \mathbf{A}(\mathbf{d}_t)^T \mathbf{X}_t + \mathbf{n}_{t+1},$$

where $\mathbf{A}(\mathbf{d}_t)$ is the state transition probability matrix with element $a_{ij}(\mathbf{d}_t) \equiv P[\mathbf{X}_{t+1} = \mathbf{e}_j | \mathbf{X}_{[t-\mathbf{1}^T\mathbf{d}_t+1:t} = \mathbf{e}_i] = a_{ij}(d_t)$, and $\mathbf{n}_{t+1} = \mathbf{X}_{t+1} - \mathbf{A}(\mathbf{d}_t)^T \mathbf{X}_t$ is a martingale increment subject to $E[\mathbf{n}_{t+1} | \mathbf{X}_{1:t}] = 0$. The observation at time t is

$$\mathbf{Y}_t = \mathbf{B}^T \mathbf{X}_t + \mathbf{w}_t,$$

where $\mathbf{w}_t = \mathbf{Y}_t - \mathbf{B}^T \mathbf{X}_t$ models the "error" subject to $E[\mathbf{w}_t | \mathbf{X}_{1:t}] = 0$.

Define the forward variables $\alpha_t(i) \equiv P[\mathbf{X}_t = \mathbf{e}_i, \mathbf{Y}_{1:t}]$ and $\boldsymbol{\alpha}_t \equiv (\alpha_t(1), \ldots, \alpha_t(M))^T$. Using the forward algorithm of Algorithm 5.2, $\boldsymbol{\alpha}_t$ can be determined. Let $\boldsymbol{\alpha}_t / P[\mathbf{Y}_{1:t}]$ be an estimation of $E[\mathbf{X}_t | \mathbf{Y}_{1:t}]$ and $\hat{\mathbf{d}}_t = diag(\boldsymbol{\alpha}_t / \mathbf{1}^T \boldsymbol{\alpha}_t)(\mathbf{d}_{t-1} + \mathbf{1})$ be an estimation of $\mathbf{d}_t$, where $P[\mathbf{Y}_{1:t}] = \sum_i \alpha_t(i) = \mathbf{1}^T \boldsymbol{\alpha}_t$. Then the forward algorithm can be approximately expressed by

$$\boldsymbol{\alpha}_{t+1} = Diag(\mathbf{B}\mathbf{Y}_{t+1}) \mathbf{A}(\hat{\mathbf{d}}_t)^T \boldsymbol{\alpha}_t$$

and

$$\hat{\mathbf{d}}_{t+1} = Diag\left(\frac{\boldsymbol{\alpha}_{t+1}}{\mathbf{1}^T \boldsymbol{\alpha}_{t+1}}\right)(\hat{\mathbf{d}}_t + \mathbf{1}),$$

where $Diag(\mathbf{B}\mathbf{Y}_{t+1})$ is an $M \times M$ diagonal matrix with the diagonal elements being observation probabilities of $\mathbf{Y}_{t+1}$ for different given states.

Define the log-likelihood of the observations up to time t given λ (Azimi et al., 2003, 2005) by

$$\begin{aligned} l_t(\lambda) &\equiv \log P[\mathbf{Y}_{1:t} | \lambda] = \sum_{\tau=1}^{t} \log P[\mathbf{Y}_\tau | \mathbf{Y}_{1:\tau-1}, \lambda] \\ &= \sum_{\tau=1}^{t} \log \frac{\mathbf{1}^T \boldsymbol{\alpha}_\tau}{\mathbf{1}^T \boldsymbol{\alpha}_{\tau-1}}, \end{aligned}$$

where $P[\mathbf{Y}_1 | \mathbf{Y}_{1:0}, \lambda]$ represents $P[\mathbf{Y}_1 | \lambda]$, and $\mathbf{1}^T \boldsymbol{\alpha}_0 = 1$.

Then the set of model parameters λ can be updated online by maximizing $l_t(\lambda)$. A sequential learning of HMM state duration using quasi-Bayes estimation was presented by Chien and Huang (2003, 2004), in which the Gaussian, Poisson, and Gamma distributions were investigated to characterize the duration models.

8.6 INFINITE HSMM AND HDP-HSMM

When the total number, M, of hidden states is unknown, extremely large or infinite, the forward–backward algorithms for the HSMMs become inapplicable, and the model selection methods based on AIC (Akaike, 1974) or BIC (Schwarz, 1978) by trying various model settings become too expensive or unpractical. For example, in the application of speaker diarization, how many speakers present, when they speak, and what characteristics governing their speech patterns are required to infer from a given single audio recording of a meeting (Fox, 2009). To address this issue, a nonparametric HSMM is extended to the HDP-HSMM (Johnson and Willsky, 2013; Nagasaka et al., 2014), which provides a powerful framework for inferring arbitrarily large state complexity from data (Beal et al., 2002), without the need to set the number of hidden states *a priori*. Analog to the generative procedure of an HSMM, the HDP-HSMM uses the sampling methodology to generate samples (states) from a set of Dirichlet process distributions or desired pdfs, and use the obtained samples for various types of inference.

Let A_i be the transition distribution from state i. Then the process HDP-HSMM can be written (Johnson and Willsky, 2013), for $i = 1, 2, \ldots, t = 1, 2, \ldots, T$,

$$\kappa \sim \mathrm{GEM}(\rho); A_i \overset{iid}{\sim} \mathrm{DP}(\iota, \kappa); S_{[t} \sim A_{S_{t-1]}};$$

$$\theta_i \overset{iid}{\sim} G; d_t \sim p_{S_t}(d; \theta_{S_t});$$

$$\varphi_i \sim H; o_{t,t+d_t-1} \sim b_{S_t,d_t}(v_{1:d_t}; \varphi_{S_t});$$

where κ is a base measure drawing from the stick-breaking distribution GEM(ρ) with parameter ρ (Fox, 2009). The transition distribution A_i draws from the Dirichlet process distribution DP(ι, κ) with concentration parameter ι and base measure κ. The state $S_{[t}$ is determined by the previous state $S_{t-1]}$ and draws from the distribution $A_{S_{t-1]}}$. θ_i is the parameter of the duration distribution of state i drawing from the duration parameter prior to distribution G. d_t is the duration of state $S_{[t}$ drawing from the duration distribution $p_{S_t}(d;\theta_{S_t})$ with parameter θ_{S_t}. φ_i is the parameter of the observation distribution of state i drawing from the observation parameter prior to distribution H. The observation probability of $o_{t:t+d_t-1}$ is determined using $b_{S_t,d_t}(o_{t:t+d_t-1};\varphi_{S_t})$ with parameter φ_{S_t}. The parameters ι and ρ can be estimated from the observations.

Denote M as the number of instantiated states, $n_{i,j}$ the number of transitions from state i to state j in the state sequence $s_{1:T}$, and $n_{i,\cdot}=\sum_j n_{i,j}$. Then the direct HDP-HSMM sampling procedure (Johnson and Willsky, 2013) includes the following main steps:

1. Given the state assignments $s_{1:T}^{(0)}$ and global transition distribution $\kappa^{(0)}\sim\mathrm{GEM}(\rho)$; let $M=\left|\bigcup_{t=1}^{T}\{s_t^{(0)}\}\right|$; $n=1$;
2. Set $s_{1:T}=s_{1:T}^{(n-1)}$ and $\kappa=\kappa^{(n-1)}$; and let $t=1$;
3. For $i=1,\ldots,M+1$:
   draw θ_i from G and φ_i from H;
   draw $s_{[t:t+d-1]}$, for $t+d-1\le T$, from

$$P[s_{[t:t+d-1]}=i|s_{1:t-1},s_{t+d:T},\kappa,H,G]$$
$$\propto\underbrace{\frac{\iota\kappa_i+n_{s_{t-1},i}}{\iota(1-\kappa_{s_{t-1}})+n_{s_{t-1},\cdot}}}_{\text{left-transition}}\cdot p_i(d;\theta_i)\cdot b_{i,d}(o_{t:t+d-1};\varphi_i)\cdot\underbrace{\frac{\iota\kappa_{s_{t+d}}+n_{i,s_{t+d}}}{\iota(1-\kappa_i)+n_{i,\cdot}}}_{\text{right-transition}};$$

4. If $s_{[t:t-d+1]}=M+1$, increment M; Let $t\leftarrow t+d$;
5. If $t\le T$, go to step 3;
6. If there exists a j such that $n_{j,\cdot}=0$ and $n_{\cdot,j}=0$, remove j and decrement M;
7. Fix $s_{1:T}^{(n)}=s_{1:T}$; Sample the global transition distribution by $\kappa^{(n)}\sim\mathrm{GEM}(\rho)$;
8. Increment n; go to step 2.

Other approaches that are able to work with high-dimensional and complex models, such as MCMC sampling or Gibbs sampling (Djuric and Chun, 2002; Economou, 2014) and beam-sampling (Dewar et al., 2012), can also be used for generating samples from a desired probability distribution function or a target distribution.

8.7 HSMM VERSUS HMM

HSMM can be considered an extension of HMM. Therefore, the HMM can be reversely considered a special case of the HSMM, in considering that the duration distributions of the HMM states are implicitly geometric. This subsection discusses the relationship between HSMM and conventional HMM. A discussion about both HMM and HSMM can also be found by Kobayashi and Yu (2007).

A hybrid HMM/HSMM proposed by Guedon (2005, 2007) can be viewed as a special case of the HSMM where the occupancy distributions of some states are constrained to be geometric distributions while others are still generally distributed.

8.7.1 HMM Using HSMM Algorithms

The HSMM usually assumes that an ended state cannot transit to itself immediately. For example, the explicit duration HSMM assumes $a_{ii} = 0$, for all $i \in \mathbf{S}$. Therefore, the relationship between an HSMM and HMM is not straightforward. Since the duration of an HMM state is implicitly a geometric distribution, we let the HSMM's duration distributions be $p_j(d) = a'^{d-1}_{jj}(1 - a'_{jj})$, with $D = \infty$, where a'_{jj} is the parameter for the geometric duration distribution of state j. Then Eqn (5.1) for the explicit duration HSMM, the most popular of the conventional HSMMs, becomes

$$\begin{aligned}
\alpha_t(j) &= \sum_{d \geq 1} \alpha^*_{t-d+1}(j) p_j(d) u_t(j,d) \\
&= p_j(1)\alpha^*_t(j) u_t(j,1) + \sum_{d \geq 2} \alpha^*_{t-d+1}(j) p_j(d) u_t(j,d) \\
&= p_j(1)\alpha^*_t(j) b_j(o_t) + a'_{jj} \sum_{d \geq 2} \alpha^*_{t-d+1}(j) p_j(d-1) u_{t-1}(j,d-1) b_j(o_t) \\
&= b_j(o_t)[(1 - a'_{jj})\alpha^*_t(j) + a'_{jj}\alpha_{t-1}(j)],
\end{aligned} \tag{8.19}$$

where $u_t(j,1) = b_j(o_t)$, $u_t(j,d) = u_{t-1}(j,d-1) b_j(o_t)$ as given by Eqn (5.10), $p_j(1) = 1 - a'_{jj}$, $p_j(d) = a'_{jj} p_j(d-1)$, and $\alpha^*_t(j) = \sum_{i \in \mathbf{S} \setminus \{j\}} \alpha_{t-1}(i) a_{ij}$ as given by Eqn (5.2). Similarly, Eqn (5.3) for the explicit duration HSMM becomes

$$
\begin{aligned}
\beta_t^*(j) &= \sum_{d\geq 1} p_j(d)u_{t+d-1}(j,d)\beta_{t+d-1}(j) \\
&= p_j(1)u_t(j,1)\beta_t(j) + \sum_{d\geq 2} p_j(d)u_{t+d-1}(j,d)\beta_{t+d-1}(j) \\
&= p_j(1)b_j(o_t)\beta_t(j) + a'_{jj}\sum_{d\geq 2} p_j(d-1)u_{t+d-1}(j,d-1)b_j(o_t)\beta_{t+d-1}(j) \\
&= b_j(o_t)[(1-a'_{jj})\beta_t(j) + a'_{jj}\beta^*_{t+1}(j)],
\end{aligned}
\tag{8.20}
$$

where $u_{t+d-1}(j,d) = u_{t+d-1}(j,d-1)b_j(o_t)$ as given by Eqn (5.9), and $\beta_t(j) = \sum_{i\in S\setminus\{j\}} a_{ji}\beta^*_{t+1}(i)$ as given by Eqn (5.4).

Equations (8.19) and (8.20) present a different forward–backward algorithm from the traditional Baum–Welch algorithm given by Eqns (1.1) and (1.2) for the HMM. The former is symmetric but the latter is not. Therefore, some applications may benefit from this symmetric property. Let $\alpha'_t(j) = \alpha_t(j)/(1-a'_{jj})$, and $a'_{ij} = (1-a'_{ii})a_{ij}$ for $i \neq j$. Then they become the same as Eqns (1.1) and (1.2) for the HMM, where a'_{ij} are the state transition probabilities of the HMM.

Inputting $p_j(d) = a'^{\,d-1}_{jj}(1-a'_{jj})$ into Eqns (5.13) and (5.14) for the variable transition HSMM, another one of the conventional HSMMs, we have

$$
a_{jj}(d) = \frac{\sum_{h\geq d+1} p_i(h)}{\sum_{h\geq d} p_i(h)} = a'_{jj},
$$

$$
a_{ij}(d) = [1-a_{ii}(d)]a_{ij} = (1-a'_{ii})a_{ij} = a'_{ij}, \quad i \neq j.
$$

Let $\alpha_t(j) = \sum_d \widehat{\alpha}_t(j,d)$ and $\beta_t(j) = \widehat{\beta}_t(j,d)$, for any d. Then the two Eqns (5.11) and (5.12) for the variable transition HSMM become the same as Eqns (1.1) and (1.2) for the HMM. The relationship between the HMM and the residual time HSMM can be similarly analyzed.

8.7.2 HSMM Using HMM Algorithms

To benefit from both the small computational complexity of HMM algorithms and the explicit duration expression of HSMMs, one often uses HMM algorithms to estimate HSMM parameters in some application areas, such as speech recognition.

The ordinary Viterbi algorithm or the Baum–Welch algorithm of HMM are usually used to find the best state/segment sequence, $(i_1^*, i_2^*, \ldots, i_T^*) = ((j_1^*, d_1^*), \ldots, (j_N^*, d_N^*))$, for given HMM parameters $\{\pi_i, a_{ij}, b_i(v_k)\}$, where $\sum_{n=1}^{N} d_n^* = T$, N is the number of segments, $i_t^*, j_n^* \in \mathbf{S}$, and $d_n^* \in \mathbf{D}$. Then the model parameters $\hat{\pi}_i$, $\hat{a}_{ij}$, $\hat{b}_i(v_k)$, and $\hat{p}_i(d)$ for the HSMM can be approximately estimated by the statistics (Chien and Huang, 2003)

$$\begin{aligned} \eta_t(i,d) &= P[S_{[t-d+1:t]} = i, o_{1:T} | \lambda] \\ &\approx I(j_n^* = i) \cdot I(d_n^* = d) \cdot I\left(\sum_{k=1}^{n} d_k^* = t\right), \end{aligned} \tag{8.21}$$

$$\begin{aligned} \xi_t(i,j) &= P[S_{t]} = i, S_{[t+1} = j, o_{1:T} | \lambda] \\ &\approx I(i_t^* = i) \cdot I(i_{t+1}^* = j), \end{aligned} \tag{8.22}$$

and

$$\begin{aligned} \gamma_t(i) &= P[S_t = i, o_{1:T} | \lambda] \\ &\approx I(i_t^* = i), \end{aligned} \tag{8.23}$$

where $I(x) = 1$ if x is true and 0 otherwise. Then the model parameters $\hat{a}_{ij}$, $\hat{b}_i(v_k)$, $\hat{p}_i(d)$, and $\hat{\pi}_i$ can be estimated by Eqns (3.11)–(3.14).

The estimated duration probabilities $\hat{p}_i(d)$ can be applied to modify the scores in the Viterbi algorithm on each departure from a state. This approach can be ensured that the resulting state segmentation sequence is more reasonable according to the duration specifications. That is, the distance metric used in the Viterbi algorithm is modified as (Rabiner, 1989)

$$\delta_t(j) = \max_{d \in \mathbf{D}} \max_{i \in \mathbf{S}} \left\{ \delta_{t-d}(i) a_{ij} [\hat{p}_j(d)]^r \prod_{\tau=t-d+1}^{t} b_j(o_\tau) \right\}, \tag{8.24}$$

for $2 \le t \le T$, $j \in \mathbf{S}$, where r is a modification factor, which is usually assumed as 1. A similar modification to the forward algorithm of HMM was given by Hanazawa et al. (1990).

In fact, Eqn (8.24) is the Viterbi HSMM algorithm for the explicit duration HSMM when $r = 1$. Generally, the best state sequence, $(i_1^*, i_2^*, \ldots, i_T^*) = ((j_1^*, d_1^*), \ldots, (j_N^*, d_N^*))$, can be found using the Viterbi HSMM algorithm of Algorithm 2.4 (Park et al., 1996; Yoma et al., 2001; Yoma and Sanchez, 2002), and then Eqns (8.21)–(8.23) are used to get the statistics for updating the model parameters.

CHAPTER 9

Applications of HSMMs

This chapter introduces major application areas in recent years.

The areas that HSMMs have been applied include, such as:

Speech synthesis	Moore and Savic (2004); Zen et al. (2004, 2007); Tachibana et al. (2005, 2006, 2008); Yamagishi and Kobayashi (2005, 2007); Yamagishi et al. (2006); Nose et al. (2006, 2007a,b); Wu et al. (2006); Schabus et al. (2014); Valentini-Botinhao et al. (2014); Maeno et al. (2014); Khorram et al. (2015)
Speech recognition	Russell and Moore (1985); Levinson (1986a); Codogno and Fissore (1987); Nakagawa and Hashimoto (1988); Gu et al. (1991); Hieronymus et al. (1992); Ratnayake et al. (1992); Chen et al. (2006); Oura et al. (2006); Pikrakis et al. (2006)
Music modeling	Liu et al. (2008); Cuvillier and Cont (2014)
Machine translation	Bansal et al. (2011)
Language identification	Marcheret and Savic (1997)
Human activity recognition	Yu et al. (2000); Mark and Zaidi (2002); Yu and Kobayashi (2003b); Hongeng and Nevatia (2003); Hongeng et al. (2004); Niwase et al. (2005); Duong et al. (2005b, 2006); Marhasev et al. (2006); Pavel et al. (2006); Zhang et al. (2006, 2008); Natarajan and Nevatia (2007a,b, 2008, 2013); Chung and Liu (2008); Boussemart and Cummings (2011); Doki et al. (2013); Park and Chung (2014); Yurur et al. (2014)
Animal activity modeling	O'Connell et al. (2011); Eldar et al. (2011); Langrock et al. (2012); Joo et al. (2013); Choquet et al. (2011, 2013)
Network traffic characterization and anomaly detection	Leland et al. (1994); Park et al. (1997); Tuan and Park (1999); Riska et al. (2002); Yu et al. (2002); Yu (2005); Li and Yu (2006); Lu and Yu (2006a); Tan and Xi (2008); Xie and Yu (2006a,b); Xie and Tang (2012); Xie et al. (2013a,b); Xu et al. (2013); Ju and Xu (2013)
Network performance	Lin et al. (2002); Wang et al. (2011); Meshkova et al. (2011); Nguyen and Roughan (2013)
Functional MRI brain mapping	Thoraval et al. (1992, 1994); Thoraval (2002); Faisan et al. (2002, 2005); Anderson et al. (2010, 2012a,b); Anderson and Fincham (2013)
Deep brain stimulation	Taghva (2011)
EEG/ECG data analysis	Thoraval et al. (1992); Hughes et al. (2003); Hughes et al. (2004); Dumont et al. (2008); McFarland et al. (2011); Oliver et al. (2012); Borst and Anderson (2015)
Early-detection of pathological events	Yang et al. (2006); Tokdar et al. (2010); Altuve et al. (2011, 2012)

(*Continued*)

(Continued)	
Equipment prognosis	Bechhoefer et al. (2006); Dong et al. (2006); He et al. (2006); Dong and He (2007a,b); Dong (2008); Wu et al. (2010); Zhao et al. (2010); Chen and Jiang (2011); Dong and Peng (2011); Peng and Dong (2011); Geramifard et al. (2011); Moghaddass and Zuo (2012a,b); Liu et al. (2012); Jiang et al. (2012); Su and Shen (2013); Boukra and Lebaroud (2014); Wang et al. (2014); Moghaddass and Zuo (2014); Wong et al. (2014); Andreoli (2014)
Diagnosis of electric power system	Zhao et al. (2014)
Document image classification	Hu et al. (1999, 2000)
Change/end-point detection for semiconductor manufacturing	Ge and Smyth (2000a,b)
Event recognition in videos	Hongeng and Nevatia (2003)
Image recognition	Takahashi et al. (2010); Liang et al. (2011); Makino et al. (2013)
Image segmentation	Lanchantin and Pieczynski (2004); Bouyahia et al. (2008); Wang et al. (2008)
Handwritten/printed text recognition	Chen et al. (1993a,b, 1995); Chen and Kundu (1994); Amara and Belaid (1996); Senior et al. (1996); Kundu et al. (1997, 1998); Cai and Liu (1998); Bippus and Margner (1999); Benouareth et al. (2008); Lin and Liao (2011)
Climate model	Sansom and Thomson (2001, 2008); Alasseur et al. (2004); Henke et al. (2012)
Icing load prognosis	Wu et al. (2014)
Prediction of particulate matter in the air	Dong et al. (2009)
Irrigation behavior	Andriyas and McKee (2014)
Dynamics of geyser	Langrock (2012)
Recognition of human genes in DNA	Kulp et al. (1996); Burge and Karlin (1997); Borodovsky and Lukashin (1998); Ter-Hovhannisyan (2008); Xi et al. (2010)
Protein structure prediction	Schmidler et al. (2000); Aydin et al. (2006); Bae et al. (2008)
Branching and flowering patterns in plants	Guedon et al. (2001); Guedon (2003, 2005, 2007); Chaubert-Pereira et al. (2010); Taugourdeau et al. (2015)
Ground target tracking	Ke and Llinas (1999)
Remote sensing	Pieczynski (2007)
Anomaly detection of spacecraft	Tagawa et al. (2011); Melnyk et al. (2013)
Terrain modeling	Wellington et al. (2005)
Prediction of earthquake	Beyreuther and Wassermann (2011)
Financial time series modeling	Bulla and Bulla (2006); Gillman et al. (2014)
Robot learning	Squire (2004); Calinon et al. (2011); Zappi et al. (2012)
Symbolic plan recognition	Duong et al. (2005a)

9.1 SPEECH SYNTHESIS

A speech synthesis system is to convert a given text to a good acoustic speech waveform that sounds natural. The fundamental difference of a synthesis system from a recognition system is that the sequence of observations (i.e., parameters for control the synthesis procedure) is to be estimated for given model parameters.

9.1.1 Speech Synthesis and ML Estimation of Observations

We first introduce concepts of speech synthesis. A waveform of speech is a continuous signal as shown in Figure 9.1(a). A speech signal is usually sampled at a rate higher than 16 kHz and windowed with a 5 or 10 ms shift. It is quite variable. Speakers with different gender, age, huskiness, and nasality usually have different accent, voice quality, mannerisms, speed, and prosody. Even for the same speaker, he/she may vary his/her word duration enormously, eliminate vowels, and change speaking style, in different context, role, and emotion. Therefore, speech signal is not appropriate for being used in training the models.

Instead, acoustic features represent less variable part of the speech signal. They are suitable for training the model or recognizing the speech. The acoustic features contain sufficient information to generate speech, including string of phones, prosody, duration, and stress/accent values for each phone, fundamental frequency for entire utterance, as well as their dynamic features of the first- and second-order differences.

Among the acoustic features, the fundamental frequency (F0) extracted from the speech signal is the lowest component of the spectrum of the speech, and its log is loosely related to pitch of the speech, as shown in Figure 9.1(b). Its dynamic features of the first- and second-order differences, that is, their delta and delta–delta are usually included in the feature vector.

The Mel-frequency cepstrum (MFC) is another acoustic feature, which is a linear cosine transform of a log power spectrum of the speech signal on a nonlinear Mel scale of frequency, where the Mel scale is used to make the features match more closely what humans hear. As an example, the first two of MFC coefficients extracted from the speech signal are shown in Figure 9.1(c).

To represent the relative energy distribution of aperiodic components, ratio between the lower and upper smoothed spectral envelopes is used

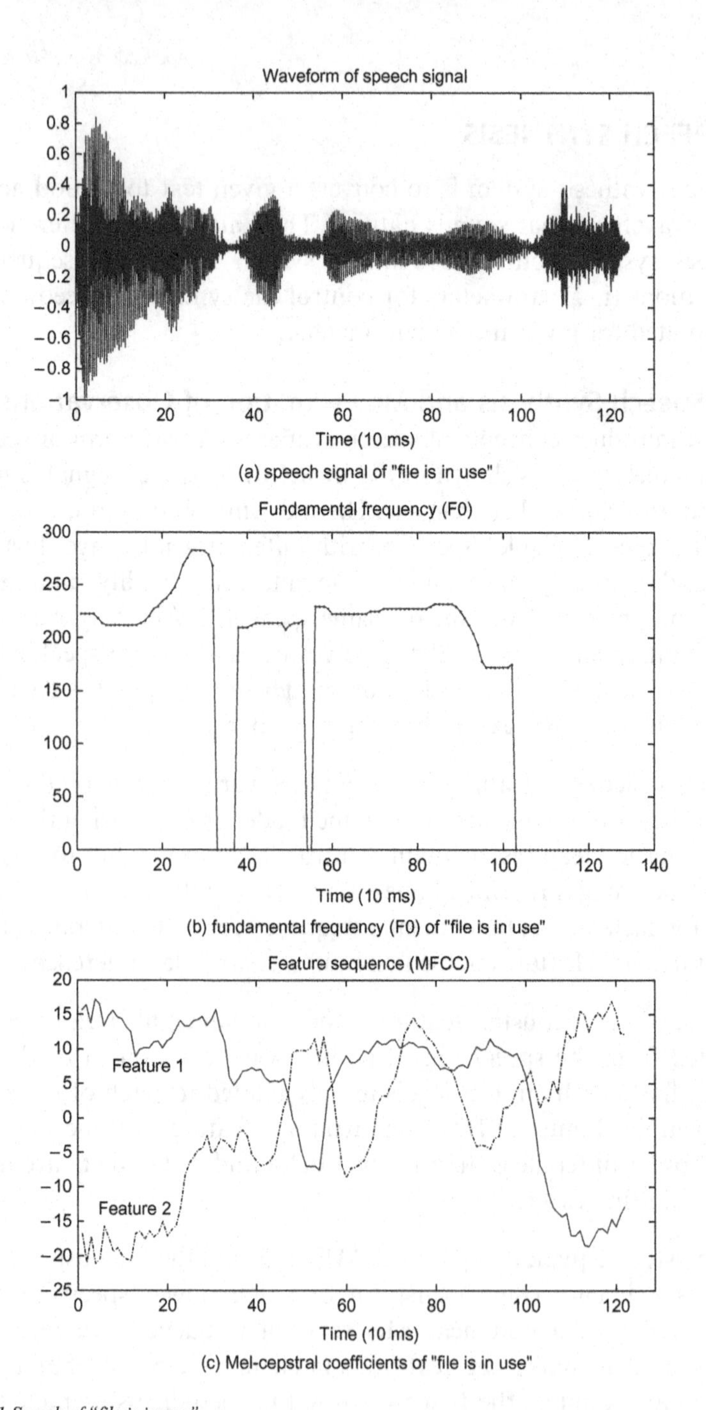

Figure 9.1 Speech of "file is in use."
(a) The waveform of the signal while a man speaks "file is in use." The length of the speech is about 1.2 s. (b) Through spectrum analysis, the fundamental frequency (F0) can be obtained, which can loosely represent the pitch of the speech. (c) The Mel-cepstral coefficients can be obtained through a linear cosine transform of the log power spectrum of the speech signal. A nonlinear Mel scale of frequency is used to make the features match more closely what humans hear. For example, two of the Mel-cepstral coefficients are presented.

as another acoustic feature, that is, aperiodicity measure. Average values of the aperiodicity measures on different frequency bands and their delta and delta–delta are used as the vector of aperiodicity measures. For example, five frequency bands, 0–1 kHz, 1–2 kHz, 2–4 kHz, 4–6 kHz, and 6–8 kHz were used by Zen et al. (2007).

All the sequences of feature vectors extracted from one speech signal are combined into a sequence of observation vectors, and used to train the context-dependent HSMMs. For example, Valentini-Botinhao et al. (2014) extracted the acoustic features that includes 59 Mel-cepstral coefficients, Mel scale F0, and 25 aperiodicity energy bands. Static, delta, and delta–delta values of the features are used as observation vectors.

A speech unit can be a subword unit (e.g., half-phone, phone, diphone, syllable, and acoustic unit), a word unit, or a group of words. Each unit can be characterized by a context-dependent HSMM. The HSMM states can be considered as the distinct sounds (e.g., phonemes and syllables) or the phases of the unit. For a large vocabulary speech synthesis, subwords are usually used as speech units. For example, Zen et al. (2007) and Maeno et al. (2014) used a five state left-to-right HSMM with no skip topology as the model of a phoneme.

Although Mel-cepstral coefficients shown in Figure 9.1(c) and aperiodicity measures can be modeled by continuous observation distributions, F0 shown in Figure 9.1(b) cannot be modeled as continuous or discrete observation distributions. Therefore, values of log F0 are usually modeled by multispace probability distributions of observations. For example, the spectrum and aperiodicity streams were modeled as single multivariate Gaussian distributions with diagonal covariance matrices (Zen et al., 2007). The F0 stream was modeled as a multispace probability distribution consisting of a Gaussian distribution (voiced space) and a probability mass function of one symbol (unvoiced space).

The duration distribution of a state is usually assumed as a parametric one such as Gaussian distribution (Oura et al., 2006), Poisson distribution (Russell and Moore, 1985), and gamma distribution (Levinson, 1986a; Codogno and Fissore, 1987). The uniform distribution with lower and upper bounding parameters was also applied, which inhibited a state occupying too few or too many speech frames (Gu et al., 1991).

There are many contextual factors (e.g., phone identity factors, stress-related factors, location factors, and part-of-speech) that affect the acoustic features. The inclusion relationship: utterance > breath phrases > accentual phrases > morae > phonemes is taken into account in considering the contextual factors. For example, Zen et al. (2007) used the number, the position, the preceding and succeeding units of phonemes, morae, accentual phrases, and breath phrases as the contextual factors.

Both the contextual factors and the acoustic features are used for training the context-dependent HSMMs. However, there is exponentially increasing number of combinations of contextual factors. All possible context-dependent model parameters cannot be estimated with a limited amount of training data. Therefore clustering the samples of contextual factors and the corresponding acoustic features into several classes and training the model parameters for those classes become necessary. In this way, new values of contextual factors or new combination of contextual factors can be classified into one or more of the known classes, and the corresponding model parameters can be used as an approximation of the model parameters for the new contextual factors. Because different acoustic features have their own influential contextual factors, they are clustered separately using different decision trees.

A speech synthesis system consists of a synthesis part and a training part, as shown in Figure 9.2.

In the training part, a lot of sentences of data with corresponding texts from one or more speakers are collected. Each of the sentence data is a waveform of speech signal. A series of acoustic features from the speech

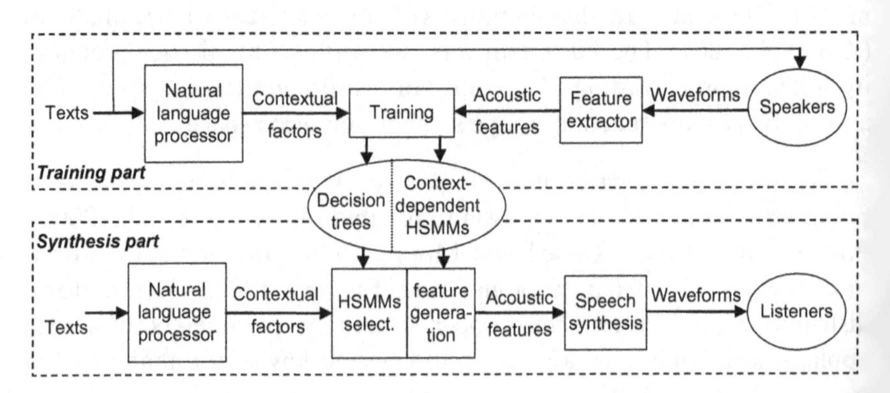

Figure 9.2 Speech synthesis system.

signal are extracted through a spectral and/or temporal analysis. In the meanwhile, the corresponding text is analyzed by the natural language processor and a series of contextual factors are obtained. Dependent on the sequences of contextual factor vectors, context-dependent HSMMs are trained from the sequences of acoustic feature vectors.

In the synthesis part, input text is analyzed and converted into linguistic specification, that is, a sequence of subwords annotated with contextual factors. Based on the sequence of subwords and the series of contextual factor vectors, corresponding context-dependent HSMMs are selected and concatenated. While retrieving the model parameters of the context-dependent HSMMs, the decision trees corresponding to different acoustic features are used to classify the contextual factors. Then a sequence of acoustic feature vectors is generated based on the model parameters of the context-dependent HSMMs. Finally, the sequence of acoustic feature vectors is used to generate a waveform of speech.

The sequence of acoustic features is generated by the ML estimation or the MAP probability given the concatenated context-dependent HSMMs. Let λ be the set of parameters of the concatenated context-dependent HSMMs, and $\mathbf{o}_{1:T}$ be the sequence of acoustic feature vectors. Then the estimation of $\mathbf{o}_{1:T}$ is to find

$$\max_{\mathbf{o}_{1:T}} P[\mathbf{o}_{1:T}|\lambda].$$

This is similar to $\max_{\lambda} P[o_{1:T}|\lambda]$. Therefore, we can define an auxiliary function similar to Eqn (3.2) by

$$Q(\mathbf{o}_{1:T},\mathbf{o}'_{1:T}) = \sum_{S_{1:T}} P[S_{1:T},\mathbf{o}_{1:T}|\lambda]\log P[S_{1:T},\mathbf{o}'_{1:T}|\lambda],$$

where $\mathbf{o}_{1:T}$ is given and $\mathbf{o}'_{1:T}$ is to be found/estimated. Then, following the similar procedure of proving Theorem 3.1, we can get an estimation formula similar to Eqn (3.10) as

$$\max_{\mathbf{o}'_{1:T}} \sum_{(j,d)} \sum_{t=d}^{T} \frac{\eta_t(j,d)}{P[\mathbf{o}_{1:T}|\lambda]} \log \frac{b_{j,d}(\mathbf{o}'_{t-d+1:t})}{b_{j,d}(\mathbf{o}_{t-d+1:t})}, \tag{9.1}$$

where the probabilities $\eta_t(j,d) \equiv P[S_{[t-d+1:t]}=j,\mathbf{o}_{1:T}|\lambda] = \alpha_t(j,d)\beta_t(j,d)$ are determined by Eqn (2.10), using a forward–backward algorithm of HSMM, and $b_{j,d}(\mathbf{o}_{t-d+1:t})$ are observation density functions. Under the assumption of conditional independency of observations for given state, that is, $b_{j,d}(\mathbf{o}_{t-d+1:t}) = \prod_{\tau=t-d+1}^{t} b_j(\mathbf{o}_\tau)$, Eqn (9.1) reduces to

$$\max_{\mathbf{o}'_{1:T}} \sum_j \sum_{t=1}^{T} \gamma_t(j) \log b_j(\mathbf{o}'_t), \tag{9.2}$$

where the probabilities $\gamma_t(j) = P[S_t = j, \mathbf{o}_{1:T}|\lambda]$ are determined by Eqn (2.13), using a forward–backward algorithm of HSMM. Suppose $b_j(\mathbf{o}'_t)$ are multispace probability distributions or a mixture of distributions as defined by Eqn (7.6),

$$b_j(\mathbf{o}'_t) = \sum_{n=1}^{N_j} p_{j,n} f_{j,n}(\mathbf{o}'_t). \tag{9.3}$$

Then similar to Eqn (7.7), we get

$$\max_{\mathbf{o}'_{1:T}} \sum_j \sum_n \sum_{t=1}^{T} \gamma_t(j,n) \log(p_{j,n} f_{j,n}(\mathbf{o}'_t)), \tag{9.4}$$

where $\gamma_t(j,n) = P[\Omega_t = (j,n), \mathbf{o}_{1:T}|\lambda] = \gamma_t(j) \frac{p_{j,n} f_{j,n}(\mathbf{o}_t)}{b_j(\mathbf{o}_t)}$. If $f_{j,n}(\mathbf{o}'_t)$ is a multivariate Gaussian distribution with the mean vector $\boldsymbol{\mu}_{j,n}$ and covariance matrix $\boldsymbol{\Sigma}_{j,n}$, Eqn (9.4) becomes

$$\max_{\mathbf{o}'_{1:T}} \sum_{t=1}^{T} \sum_j \sum_n \gamma_t(j,n) \left(-\frac{1}{2} (\mathbf{o}'_t - \boldsymbol{\mu}_{j,n})^T \boldsymbol{\Sigma}_{j,n}^{-1} (\mathbf{o}'_t - \boldsymbol{\mu}_{j,n}) \right) + C,$$

where C is a constant independent of $\mathbf{o}'_{1:T}$, and the superscript T represents transpose of the column vector. Eqn (9.4) can then be maximized by letting its partial derivative for $\mathbf{o}'_t$ be zeros, that is,

$$\sum_j \sum_n \gamma_t(j,n)(-\mathbf{e}^T \boldsymbol{\Sigma}_{j,n}^{-1} (\mathbf{o}'_t - \boldsymbol{\mu}_{j,n})) = 0$$

or

$$\mathbf{e}^T \left(\sum_j \sum_n \gamma_t(j,n) \boldsymbol{\Sigma}_{j,n}^{-1} \right) \mathbf{o}'_t = \mathbf{e}^T \sum_j \sum_n \gamma_t(j,n) \boldsymbol{\Sigma}_{j,n}^{-1} \boldsymbol{\mu}_{j,n}, \tag{9.5}$$

for $t = 1, \ldots, T$, where $\mathbf{e}^T = (1, \ldots, 1)$.

In summary, the procedure to find the sequence of acoustic feature $\mathbf{o}_{1:T}$ is

Algorithm 9.1 ML Estimation of Observations

1. *Determine the set of model parameters* λ, *including the mean vector* $\boldsymbol{\mu}_{j,n}$ *and covariance matrix* $\boldsymbol{\Sigma}_{j,n}$ *of the multivariate Gaussian distributions of observation vectors;*
2. *Set an initial sequence of* $\mathbf{o}_{1:T}$*;*
3. *Calculate* $\gamma_t(j) = P[S_t = j, \mathbf{o}_{1:T}|\lambda]$ *by* Eqn (2.13), *using the forward–backward algorithm of* Algorithm 2.1; *further determine* $\gamma_t(j,n) = \gamma_t(j)\dfrac{p_{j,n}f_{j,n}(\mathbf{o}_t)}{b_j(\mathbf{o}_t)}$*;*
4. *Solve* Eqn (9.5) *to get* $\mathbf{o}'_{1:T}$*;*
5. *Let* $\mathbf{o}_{1:T} = \mathbf{o}'_{1:T}$, *and go back to Step 3; repeat this procedure until the ML* $\max_{\mathbf{o}_{1:T}} P[\mathbf{o}_{1:T}|\lambda]$ *is reached.*

9.1.2 Other Applications Similar to Speech Synthesis

A joint audiovisual HSMM was proposed by Schabus et al. (2014). In this work, coupled with the acoustic models used for speech synthesis, visual models produce a sequence of 3D motion tracking data to be used for animating a talking head. The acoustic features, containing 39 + 1 Mel-cepstral features, log F0 and 25 band-limited aperiodicity measures, and the visual features are simultaneously used to train context-dependent HSMMs, each of which is a left-to-right HSMM with five states, multistream observations, and multispace distribution (MSD) of observations.

Speech synthesis can be applied for synthesizing a song. Park et al. (2010) constructed HSMMs to model features of song notes ranging from E3 to G5. The notes indicated by the musical score are used for determining contextual factors. Various melodies sung neutrally with restricted set of words are used for extracting acoustic features of pitch and duration. The constructed HSMMs then are applied to control the expressed emotion of a synthesized song.

Similar models and algorithms for speech synthesis can be applied to synthesis of human walking motion animation (Yamazaki et al., 2005; Niwase et al., 2005; Tilmanne and Dutoit, 2012a,b). Human motion is decomposed into motion primitives. A sequence of symbols representing motion primitives is used to describe the motion. Each motion primitive is modeled by an adaptive-factor HSMM. Then the model parameters are estimated by Eqn (8.6). In the motion synthesis part, a desired motion description is converted to a sequence of motion symbols/primitives. Corresponding to the sequence of motion symbols, HSMMs are selected and concatenated. Using the concatenated HSMMs, a sequence of human body model parameters, including their dynamic features denoted by the first and second time derivatives, is generated based on the ML estimation algorithm similar to Algorithm 9.1. The parameters are eventually visualized in 3D image and converted into motion animation.

Besides, the acoustic features and the context-dependent HSMMs can be applied for speech recognition. The related work for speech recognition can be found, such as Levinson (1986a), Codogno and Fissore (1987), Nakagawa and Hashimoto (1988), Gu et al. (1991), Ratnayake et al. (1992), Hieronymus et al. (1992), and Oura et al. (2006).

9.2 HUMAN ACTIVITY RECOGNITION

Human activity recognition (HAR) is essential in surveillance, smart environment, and man-machine conversation. To recognize human activities, external and wearable sensors are used. The external sensors, such as cameras, are used to capture user's trace, pose, salient body parts, related objects, scene context, etc. The wearable sensors or built-in smart device sensors are attached to a user to measure their location/movement, physiological signals, and environmental variables.

For the HAR using wearable sensors or built-in smart device sensors, triaxial accelerometers, and GPS are usually used for measuring attributes related to the user's movement, location, or transportation mode. Heart rate, respiration rate, skin temperature, skin conductivity, and electrocardiogram are used for physiological signals. Temperature, humidity, altitude, barometric pressure, light and audio level are used for describing individual's surrounding.

After getting multiple series of signals that are related to the user and its surrounding, feature extraction methods are applied to each time window to filter relevant information and obtain quantitative measures. Discrete Cosine Transform (DCT), Principal Component Analysis (PCA), and autoregressive model coefficients have been applied to extract time-domain and frequency-domain features of acceleration signals. Polynomials functions have been used to fit physiological signals to describe the morphological interrelationship among data.

For the HAR using external sensors, videos, image sequences, or still images are usually used to extract spatial-temporal features, bag-of-word features, and semantic features. The apparent motion of individual pixels on the image plane can track the physical motion projected onto the image plane. Moving parts of a scene can be segmented out from the background. A spatial Gaussian and a derivative filter of Gaussian on the temporal axis can show high responses at regions of motion in a video sequence. Human knowledge about intrinsic properties of human pose, salient body parts, related objects, scene context, and spatial-temporal attributes can be applied to enhance the estimation of the features. For example, the spatial-temporal patterns of arms and legs can be applied for recognizing a walking action.

Based on a sequence of feature vectors extracted from the raw signals, the temporal dynamics of a motion can be modeled. In the training stage, multiple sequences of feature vectors are collected from individuals performing each activity. Then, learning methods of HSMMs are used to generate an activity recognition model from the sequences of feature vectors. The trained HSMMs can then be used to recognize activities from a sequence of newly collected feature vectors.

As an example, in a smart home there are a given set of furniture and appliances. Spending time at one of these designated places is called an atomic activity (Duong et al., 2005a,b, 2006). For example, spending time at the cupboard, stove, fridge, etc. A major activity of human daily routine, such as making breakfast, eating breakfast, going to work, coming back home, etc., contains a sequence of atomic activities with different durations. For example, making breakfast usually contains the following sequence: walking to the fridge, spending some time there for taking out the food, going to the stove, and staying there for cooking breakfast. Duong et al. (2005a,b, 2006) introduced a two-layered extension of HSMM for modeling the activities. In this model, a sequence of major activities is modeled by a Markov chain, with each state representing a major activity. That is, the set of states is denoted by Q^* defining a Markov chain. Each given major activity/state $q^* \in Q^*$ is modeled by an HSMM, containing a sequence of atomic activities, with each HSMM state representing an atomic activity and its duration representing the atomic action time. Suppose, the set of HSMM states is Q. Since for different major activity, the transition probabilities between atomic activities and their duration distributions may be different, the model parameters of the q^*th HSMM corresponding to major activity/state $q^* \in Q^*$ is thus denoted by λ_{q^*}. The model parameters for the HSMMs are $\lambda_1, \ldots, \lambda_{|Q^*|}$. The state duration distributions of the HSMMs can be modeled by various pdfs. In Duong et al. (2005a), they were modeled by the multinomial and the discrete Coxian parameterization. The observation sequence is the locations of the person captured by several cameras, which was used to train the two-layered HSMM and recognize activities of daily living.

Similarly, Zhang et al. (2006) used a layered HSMM in an intelligent surveillance system deployed in parking lots. Chung and Liu (2008) applied a hierarchical context hidden Markov model for behavior understanding from video streams in a nursing center. A three-layered variable transition HMM was introduced by Natarajan and Nevatia (2008) for representing the composite actions at the top-most layer, the primitive actions at the middle layer and the body pose transitions at the bottom-most layer.

As an extension to the layered HSMM, Natarajan and Nevatia (2007a) proposed a hierarchical multichannel HSMM, which is used to model multiple interacting processes. As a special case of multichannel HSMM, a coupled HSMM is used for continuous sign-language recognition in Natarajan and Nevatia (2007b). These models were extended to graphic models in Natarajan and Nevatia (2013).

Except the layered and multichannel HSMMs, conventional HSMMs were also applied in activity recognition. Yu et al. (2000) and Mark and Zaidi (2002) applied an explicit duration HSMM into mobility tracking in wireless networks, which was further used to solve the problems of missing data and nonsynchronous multiple observation sequences (Yu and Kobayashi, 2003b). Xu and Jiang (2012) proposed an HSMM-based Cell-ID localization algorithm for mobility tracking. Yurur et al. (2014) used a discrete-time inhomogeneous hidden semi-Markov model (DT-IHS-MM) to describe user states of sitting, standing, walking, and running. The states are estimated from signal of accelerometer sensor with missing observations when power efficiency is taken into consideration at the low-level sensory operations Pavel et al. (2006) used an HSMM in unobtrusive assessment of mobility. Hongeng and Nevatia (2003), Hongeng et al. (2004), and Zhang et al. (2008) applied HSMMs for recognizing events in a video surveillance. Marhasev et al. (2006) used a non-stationary HSMM in activity recognition. Park and Chung (2014) used an HSMM to model the characteristics of dynamic motor control strategy of humans, based on a bio-electric signal measured on the skin surface using electromyogram. Doki et al. (2013) used a left-to-right HSMM to express time series data obtained by sensing human actions and situations around a person.

Similar to the HAR, the HSMMs are applied to animal activity recognition by O'Connell et al. (2011), Eldar et al. (2011), Langrock et al. (2012), Joo et al. (2013), and Choquet et al. (2011, 2013), and unmanned aerial vehicle (UAV) activity recognition by Zha et al. (2013).

9.3 NETWORK TRAFFIC CHARACTERIZATION AND ANOMALY DETECTION

In this application, HSMMs are applied to characterize the network traffic. Measurements of real traffic often indicate that a significant amount of variability is presented in the traffic observed over a wide range of time scales, exhibiting self-similar or long-range dependent characteristics (Leland et al., 1994). Such characteristics can have a significant impact on the performance of networks and systems (Tuan and Park, 1999; Park et al., 1997). Therefore, better understanding the nature of network traffic is critical for network design, planning, management, and security. A major advantage of using an HSMM is the capability of capturing various statistical properties of the traffic, including the long-range dependence (Yu et al., 2002). It can also be used together with, for example, matrix-analytic methods to obtain analytically tractable solutions to queueing-theoretic models of server performance (Riska et al., 2002).

In this application, an observation in the observation sequence represents the number of user requests/clicks, packets, bytes, connections, etc., arriving in a time unit. It can also be the inter-arrival time between requests, packets, URLs, or protocol keywords. The observation sequence is characterized as a discrete-time random process modulated by an underlying (hidden state) semi-Markov process. The hidden state represents the density of traffic, mass of active users, or a web page that is hyperlinked with others.

Using the HSMM trained by the normal behavior, one can detect anomaly embedded in the network behavior according to its likelihood or entropy against the model (Yu, 2005; Li and Yu, 2006; Lu and Yu, 2006a; Xie and Yu, 2006a,b; Xie and Zhang, 2012; Xie and Tang, 2012; Xie et al., 2013a,b), recognize user click patterns (Xu et al., 2013), extract users' behavior features (Ju and Xu, 2013) for SaaS (Software as a Service), or estimate the packet loss ratios and their confidence intervals (Nguyen and Roughan, 2013).

For example, a web workload (requests/s) recorded in the peak hour is shown in Figure 1.7 (gray line). The arrival process for a given state j can be assumed as, for instance, Poisson process $b_j(k) = \mu_j^k e^{-\mu_j}/k!$ with one parameter μ_j. The initial value of μ_j is assumed to be proportional to its state index j, that is,

$$\mu_j = \max(o_t) \cdot j/M,$$

so that higher state corresponds to higher arrival rate, where M is the total number of hidden states. In considering the range of the observable values (requests/s), the total number M of hidden states is initially assumed to be 30. During the re-estimation procedure, the states that are never visited will be deleted from the state space. To characterize the second order self-similar property of the workload, the duration of state j can be assumed as, for instance, a heavy-tailed Pareto distribution $p_j(d) = \lambda_j d^{-(\lambda_j+1)}$ with one parameter λ_j. The initial values of λ_j can be assumed equal for all states. To reduce the computational amount, the maximum duration D of the states can be assumed to be finite with sufficiently large value to cover the maximum duration of any state in the given observation sequence, where $D = 500$ s is assumed. As a reasonable choice, the initial values of the probabilities a_{ij} and π_j are assumed uniform.

Given these initial assumptions for the explicit duration HSMM, the ML model parameters can be estimated using the re-estimation algorithm of Algorithm 3.1. The MAP states S_t, for $t = 1, \ldots, T$, can be estimated using Eqn (2.15). The results showed that there were 20 hidden states modulating the arrival rate of requests, and only 41 state transitions occurring during 3600 s. The maximum duration D went up to 405 and the process stayed in the same state for a mean duration of 87.8 s. There were two classes of states among the 20 states: 5 states in the middle played a major role in modulating the arrival streams in the sens that the process spent most time in these 5 states; and the remaining 15 states having the higher and lower indices represented the rare situations that had ultra high or low arrival rates lasting very short time.

9.4 fMRI/EEG/ECG SIGNAL ANALYSIS

Applying an HSMM in functional magnetic resonance imaging (fMRI) is to reveal components of interest in the fMRI data, as did in Faisan et al. (2002, 2005), Thoraval (2002) and Thoraval et al. (1992, 1994). For example, Faisan et al. (2005) used the HSMM to automatically detect neural activation embedded in a given set of fMRI signals. The brain mapping is enriched with activation lag mapping, activation mode visualizing, and hemodynamic response function (HRF) analysis. The shape variability, neural event timing, and fMRI response linearity of the HRF were solved by the HSMM.

In this application, the sequence of hemodynamic response onsets (HROs) observed in the fMRI signal is used as the observation sequence. A left-to-right Markov chain is selected for the hidden process of task-induced neural activations. The inter-state duration distributions are specified as one-dimensional Gaussians. Because observation sequence is usually composed of events mixed with missing observations (null), the probability of missing an observation is assumed as $1 - h_{ij}$ and having an observation as h_{ij} during the transition from state i to state j. The probability of observing event $o_t = e_l$ is $b_{ij}(e_l)$, specified as one-dimensional Gaussians.

Anderson et al. (2010, 2012a,b) and Anderson and Fincham (2013) applied HSMMs to identify stages in fMRI data, and four stages, including encoding the problems, planning a solution strategy, solving the problems, and entering a response, were discovered. Each state is defined by a neural signature, as well as by a gamma distribution describing the state's duration over the trials in the experiment.

Compared with fMRI which has second-range temporal resolution, Electroencephalography (EEG) has millisecond-range temporal resolution. It can diagnose epilepsy, sleep disorders, coma, encephalopathies, brain death, etc. It uses multiple electrodes placed on the scalp to record electrical activity of the brain along the scalp. The measured voltage fluctuations reflect the ionic current within the neurons of the brain. HSMMs are useful models for analyzing the EEG signals.

For example, Borst and Anderson (2015) applied HSMMs for identifying processing stages in human information processing from EEG signals. EEG were recorded from 32 Ag-AgCl sintered electrodes, which created one snapshot/sample every 80 ms containing a set of 640 data. A spatial PCA was then applied to the 640 data, and the first 100 PCA components were used as a vector of extracted features/observations. The observations in each vector for given state are assumed to be independent Gaussians, because the PCA factors are basically distributed as independent normal distributions. Suppose the PCA factors for snapshot t are f_{tk}, that is, $o_t = (f_{t1}, \ldots, f_{t100})$, and the mean and variance of the Gaussian distribution for f_{tk} given state i are μ_{ik} and σ_{ik}, respectively. Then the observation distribution is

$$b_i(o_t) = \prod_{k=1}^{100} N(f_{tk}; \mu_{ik}, \sigma_{ik}),$$

where $N(f_{tk}; \mu_{ik}, \sigma_{ik})$ are Gaussian distributions. To model processing stages in human information processing, a left-to-right HSMM was assumed, with each state representing a processing stage (e.g., encoding, familiarity, recollection, and response) in the task. The brain signature, that is, μ_{ik} and σ_{ik}, of a processing stage is assumed invariable in different experimental conditions, but the state duration distribution is assumed to be a gamma distribution with different parameters. This can reflect how the experimental conditions affect the duration of a processing stage. Suppose v_{ic} and a_{ic} are the shape and scale parameters of gamma distribution for given state i and experimental condition c. Then the probability that state i stays for duration d (snapshots) under experimental condition c is

$$p_{ic}(d) = G(80d; v_{ic}, a_{ic})$$

or

$$p_{ic}(d) = \int_{80d-40}^{80d+40} G(\tau; v_{ic}, a_{ic}) d\tau,$$

where $G(\tau; v_{ic}, a_{ic})$ are gamma distributions, τ is the time in milliseconds, and $80d$ is the time length of d snapshots.

REFERENCES

Achan, K., Roweis, S., Hertzmann, A., Frey, B., 2005. A segment-based probabilistic generative model of speech. In: Proc. of the 2005 IEEE International Conference on Acoustics, Speech, and Signal Processing, vol. 5, Philadelphia, PA, USA, pp. 221–4.

Ait-el-Fquih, B., Desbouvries, F., 2005. Kalman filtering for triplet Markov chains: applications and extensions. In: Proceedings of the International Conference on Acoustics, Speech and Signal Processing, (ICASSP 05), vol. 4, Philadelphia, PA, USA, pp. 685–8.

Akaike, H., 1974. A new look at the statistical model identification. IEEE Trans. Automat. Control 19 (6), 716–723.

Alasseur, C., Husson, L., Perez-Fontan, F., 2004. Simulation of rain events time series with Markov model. In: Proc. of 15th IEEE International Symposium on Personal, Indoor and Mobile Radio Communications, (PIMRC 2004), vol. 4, pp. 2801–5.

Altuve, M., Carrault, G., Beuchee, A., Pladys, P., Hernandez, A.I., 2011. On-line apnea-bradycardia detection using hidden semi-Markov models. In: 33rd Annual International Conference of the IEEE EMBS, Boston, Massachusetts USA, August 30–September 3, 2011, pp. 4374–7.

Altuve, M., Carrault, G., Beuchee, A., Flamand, C., Pladys, P., Hernandez, A.I., 2012. Comparing hidden Markov model and hidden semi-Markov model based detectors of apnea-bradycardia episodes in preterm infants. Comput. Cardiol. 39, 389–392.

Amara, N.B., Belaid, A., 1996. Printed PAW recognition based on planar hidden Markov models. In: Proceedings of the 13th International Conference on Pattern Recognition, vol. 2, pp. 220–4.

Anderson, J.R., Fincham, J.M., 2013. Discovering the sequential structure of thought. Cogn. Sci. 37 (6), 1–31.

Anderson, J.R., Betts, S., Ferris, J.L., Fincham, J.M., 2010. Neural imaging to track mental states while using an intelligent tutoring system. Proc. Natl. Acad. Sci. U.S.A. 107 (15), 7018–7023. Available from: http://dx.doi.org/10.1073/pnas.1000942107.

Anderson, J.R., Betts, S., Ferris, J.L., Fincham, J.M., 2012a. Tracking children's mental states while solving algebra equations. Hum. Brain Mapp. 33, 2650–2665.

Anderson, J.R., Fincham, J.M., Schneider, D.W., Yang, J., 2012b. Using brain imaging to track problem solving in a complex state space. NeuroImage 60 (1), 633–643.

Andreoli, J.-M. 2014. Learning energy consumption profiles from data. In: 2014 IEEE Symposium on Computational Intelligence and Data Mining (CIDM), December 9–12, 2014, pp. 463–70.

Andriyas, S., McKee, M., 2014. Exploring irrigation behavior at Delta, Utah using hidden Markov models. Agric. Water Manag. 143 (2014), 48–58.

Ariki, Y., Jack, M.A., 1989. Enhanced time duration constraints in hidden Markov modelling for phoneme recognition. Electron. Lett. 25 (13), 824–825.

Askar, M., Derin, H., 1981. A recursive algorithm for the Bayes solution of the smoothing problem. IEEE Trans. Automat. Control AC-26, 558–561.

Austin, S.C., Fallside, F., 1988. Frame compression in hidden Markov models. In: Proc. of 1988 International Conference on Acoustics, Speech, and Signal Processing, ICASSP-88, April 11–14, 1988, pp. 477–80.

Aydin, Z., Altunbasak, Y., Borodovsky, M., 2006. Protein secondary structure prediction for a single-sequence using hidden semi-Markov models. BMC Bioinformatics 7, 178.

Azimi, M., Nasiopoulos, P., Ward, R.K., 2003. Online identification of hidden semiMarkov models. In: Proceedings of the 3rd International Symposium on Image and Signal Processing and Analysis, ISPA, vol. 2, September 18–20, 2003, pp. 991–6.

Azimi, M., Nasiopoulos, P., Ward, R.K., 2004. A new signal model and identification algorithm for hidden semi-Markov signals. In: Proceedings of IEEE International Conference on Acoustics, Speech, and Signal Processing, 2004, (ICASSP '04), vol. 2, May 17–21, 2004, pp. ii-521–4.

Azimi, M., Nasiopoulos, P., Ward, R.K., 2005. Offline and online identification of hidden semi-Markov models. IEEE Trans. Signal Process. 53 (8, Part 1), 2658–2663.

Bae, K., Mallick, B.K., Elsik, C.G., 2008. Prediction of protein interdomain linker regions by a nonstationary hidden Markov model. J Am Stat. Assoc. 103 (483), 1085–1099.

Bahl, L.R., Cocke, J., Jelinek, F., Raviv, J., 1974. Optimal decoding of linear codes for minimizing symbol error rate. IEEE Trans. Inf. Theory, Vol. 20, 284–287.

Bansal, M., Quirk, C., Moore, R.C., 2011. Gappy phrasal alignment by agreement. The 49th Annual Meeting of the Association for Computational Linguistics: Human Language Technologies, June 19–24, 2011, pp. 1308–17.

Barbu, V.S., Limnios, N., 2008. Semi-Markov chains and hidden semi-Markov models toward applications: their use in reliability and DNA analysis. Springer, New York, ISBN 978-0-387-73171-1.

Baum, L.E., Petrie, T., 1966. Statistical inference for probabilistic functions of finite state Markov chains. Ann. Math. Stat. 37, 1554–1563.

Beal, M.J., Ghahramani, Z., Rasmussen, C.E., 2002. The infinite hidden Markov model. Adv. Neural Inf. Process. Syst. 14, 577–584.

Bechhoefer, E., Bernhard, A., He, D., Banerjee, P., 2006. Use of hidden semi-Markov models in the prognostics of shaft failure. In: Proceedings American Helicopter Society 62th Annual Forum, Phoenix, AZ. Available from: <http://www.vtol.org/pdf/62se.pdf>.

Benouareth, A., Ennaji, A., Sellami, M., 2008. Arabic handwritten word recognition using HMMs with explicit state duration. EURASIP J. Adv. Signal Process. 2008, 1–13.

Beyreuther, M., Wassermann, J., 2011. Hidden semi-Markov Model based earthquake classification system using Weighted Finite-State Transducer. Process. Geophys. 18, 81–89.

Bippus, R., Margner, V., 1999. Script recognition using inhomogeneous P2DHMM and hierarchical search space reduction. In: Proceedings of the Fifth International Conference on Document Analysis and Recognition, 1999 (ICDAR '99), September 20–22, 1999, pp. 773–6.

Borst, J.P., Anderson, J.R., 2015. The discovery of processing stages: analyzing EEG data with hidden semi-Markov models. NeuroImage 108 (2015), 60–73.

Bonafonte, A., Ros, X., Marino, J.B., 1993. An efficient algorithm to find the best state sequence in HSMM. In: Proceedings of Eurospeech'93, Berlin, pp. 1547–50.

Bonafonte, A., Vidal, J., Nogueiras, A., 1996. Duration modeling with expanded HMM applied to speech recognition. In: Fourth International Conference on Spoken Language, 1996 (ICSLP 96), vol. 2, October 3–6, 1996, pp. 1097–100.

Borodovsky, M., Lukashin, A.V., 1998. GeneMark. hmm: new solutions for gene finding. Nucleic Acids Res. 26, 1107–1115.

Boukra, T., Lebaroud, A., 2014. Identifying new prognostic features for remaining useful life prediction. In: 16th International Power Electronics and Motion Control Conference and Exposition Antalya, Turkey, pp. 1216–21.

Boussemart, Y., Cummings, M.L., 2011. Predictive models of human supervisory control behavioral patterns using hidden semi-Markov models. Eng. Appl. Artif. Intell. 24 (7), 1252–1262.

Boutillon, E., Gross, W.J., Gulak, P.G., 2003. VLSI architectures for the MAP algorithm. IEEE Trans. Commun. 51 (2), 175–185.

Bouyahia, Z., Benyoussef, L., Derrode, S., 2008. Change detection in synthetic aperture radar images with a sliding hidden Markov chain model. J. Appl. Remote Sens. 2, 023526.

Bulla, J., Bulla, I., 2006. Stylized facts of financial time series and hidden semi-Markov models. Comput. Stat. Data Anal. 51 (4), 2192–2209.

Bulla, J., Bulla, I., Nenadic, O., 2010. hsmm—An R package for analyzing hidden semi-Markov models. Comput. Stat. Data Anal. 54 (2010), 611–619.

Burge, C., Karlin, S., 1997. Prediction of complete gene structures in human genomic DNA. J. Mol. Biol. 268, 78–94.

Burshtein, D., 1995. Robust parametric modeling of durations in hidden Markov models. In: Proc. of 1995 International Conference on Acoustics, Speech, and Signal Processing, (ICASSP-95), vol. 1, May 9–12, 1995, pp. 548.

Burshtein, D., 1996. Robust parametric modeling of durations in hidden Markov models. IEEE Trans. Speech Audio Process. 4 (3), 240–242.

Cai, J., Liu, Z.-Q., 1998. Integration of structural and statistical information for unconstrained handwritten numeral recognition. In: Proceedings of Fourteenth International Conference on Pattern Recognition, vol. 1, August 16–20, 1998, pp. 378–80.

Calinon, S., Pistillo, A., Caldwell, D.G., 2011. Encoding the time and space constraints of a task in explicit-duration Hidden Markov Model. In: 2011 IEEE/RSJ International Conference on Intelligent Robots and Systems, September 25–30, 2011. San Francisco, CA, USA, pp. 3413–18.

Chaubert-Pereira, F., Guedon, Y., Lavergne, C., Trottier, C., 2010. Markov and semi-Markov switching linear 1 mixed models used to identify forest tree growth components. Biometrics 66 (3), 753–762.

Chen, J.-H., Jiang, Y.-C., 2011. Development of hidden semi-Markov models for diagnosis of multiphase batch operation. Chem. Eng. Sci. 66, 1087–1099.

Chen, K., Hasegawa-Johnson, M., Cohen, A., Borys, S., Kim, S.-S., Cole, J., et al., 2006. Prosody dependent speech recognition on radio news corpus of American English. IEEE Trans. Audio Speech Lang. Process. 14 (1), 232–245 (see also IEEE Transactions on Speech and Audio Processing).

Chen, M.-Y., Kundu, A., 1994. A complement to variable duration hidden Markov model in handwritten word recognition. In: Proceedings of IEEE International Conference on Image Processing, 1994 (ICIP-94), vol. 1, November 13–16, 1994, pp. 174–8.

Chen, M.-Y., Kundu, A., Srihari, S.N., 1993a. Handwritten word recognition using continuous density variable duration hidden Markov model. In: Proc. of 1993 IEEE International Conference on Acoustics, Speech, and Signal Processing (ICASSP-93), vol. 5, April 27–30, 1993, pp. 105–8.

Chen, M.-Y., Kundu, A., Srihari, S.N., 1993b. Variable duration hidden Markov model and morphological segmentation for handwritten word recognition. In: Proc. of 1993 IEEE Computer Society Conference on Computer Vision and Pattern Recognition (CVPR '93), June 15–17, 1993, pp. 600–1.

Chen, M.Y., Kundu, A., Srihari, S.N., 1995. Variable duration hidden Markov model and morphological segmentation for handwritten word recognition. IEEE Trans. Image Process. 4, 1675–1688.

Chien, J.-T., Huang, C.-H., 2003. Bayesian learning of speech duration models. IEEE Trans. Speech Audio Process. 11 (6), 558–567.

Chien, J.-T., Huang, C.-H., 2004. Bayesian duration modeling and learning for speech recognition. In: Proc. of IEEE International Conference on Acoustics, Speech, and Signal Processing, 2004 (ICASSP '04), vol. 1, May 17–21, 2004, pp. I–1005–8.

Choquet, R., Viallefont, A., Rouan, L., Gaanoun, K., Gaillard, J.-M., 2011. A semi-Markov model to assess reliably survival patterns from birth to death in free-ranging populations. Methods Ecol. Evol. 2011 (2), 383–389.

Choquet, R., Guédon, Y., Besnard, A., Guillemain, M., Pradel, R., 2013. Estimating stop over duration in the presence of trap-effects. Ecol. Model. 250, 111–118.

Chung, P.C., Liu, C.D., 2008. A daily behavior enabled hidden Markov model for human behavior understanding. Pattern Recogn. 41, 1572–1580.

Cinlar, E., 1975. Markov renewal theory: a survey. Manage. Sci. 21 (7), 727–752, Theory Series (Mar., 1975).

Codogno, M., Fissore, L., 1987. Duration modelling in finite state automata for speech recognition and fast speaker adaptation. In: Proc. of IEEE International Conference on Acoustics, Speech, and Signal Processing (ICASSP '87), vol. 12, April 1987, pp. 1269–72.

Cohen, Y., Erell, A., Bistritz, Y., 1997. Enhancement of connected words in an extremely noisy environment. IEEE Trans. Speech Audio Process. 5 (2), 141–148.

Cuvillier, P., Cont, A., 2014. Coherent time modeling of semi-Markov models with application to real-time audio-to-score alignment. In: 2014 IEEE International Workshop on Machine Learning for Signal Processing, September 21–24, 2014, Reims, France.

Dean, T., Kanazawa, K., 1989. A model for reasoning about persistence and causation. Artif. Intell. 93 (1–2), 1–27.

Dempster, A.P., Laird, N.M., Rubin, D.B., 1977. Maximum likelihood from incomplete data via the EM algorithm. J. Royal Stat. Soc. Series B (Methodological) 39, 1–38.

Deng, L., Aksmanovic, M., 1997. Speaker-independent phonetic classification using hidden Markov models with mixtures of trend functions. IEEE Trans. Speech Audio Process. 5 (4), 319–324.

Deng, L., Aksmanovic, M., Sun, X., Wu, J., 1994. Speech recognition using hidden Markov models with polynomial regression functions as nonstationary states. IEEE Trans. Speech Audio Process. 2 (4), 507–520.

Devijver, P.A., 1985. Baum's forward-backward algorithm revisited. Pattern Recogn. Lett. 3, 369–373.

Dewar, M., Wiggins, C., Wood, F., 2012. Inference in hidden Markov models with explicit state duration distributions. IEEE Signal Process. Lett. 19 (4), 235–238.

Ding, J.-R., Shah, S.P., 2010. Robust hidden semi-Markov modeling of array CGH data. In: 2010 IEEE International Conference on Bioinformatics and Biomedicine, pp. 603–8.

Djuric, P.M., Chun, J.-H., 1999. Estimation of nonstationary hidden Markov models by MCMC sampling. In: 1999 IEEE International Conference on Acoustics, Speech, and Signal Processing (ICASSP '99), vol. 3, March 15–19, 1999, pp. 1737–40.

Djuric, P.M., Chun, J.-H., 2002. An MCMC sampling approach to estimation of nonstationary hidden Markov models. IEEE Trans. Signal Process. 50 (5), 1113–1123.

Doki, K., Hirai, T., Hashimoto, K., Doki, S., 2013. A modeling method for human actions considering their temporal and spatial differences. In: 4th IEEE International Conference on Cognitive Infocommunications, December 2–5, 2013, pp. 857–62.

Dong, M., 2008. A novel approach to equipment health management based on auto-regressive hidden semi-Markov model (AR-HSMM). Sci. China Series F Inf. Sci. 51 (9), 1291–1304.

Dong, M., He, D., 2007a. A segmental hidden semi-Markov model (HSMM)-based diagnostics and prognostics framework and methodology. Mech. Syst. Signal Process. 21 (5), 2248–2266.

Dong, M., He, D., 2007b. Hidden semi-Markov model-based methodology for multi-sensor equipment health diagnosis and prognosis. Eur. J. Oper. Res. 178 (3), 858–878.

Dong, M., Peng, Y., 2011. Equipment PHM using non-stationary segmental hidden semi-Markov model. Rob. Comput.-Integr. Manuf. 27 (2011), 581–590.

Dong, M., He, D., Banerjee, P., Keller, J., 2006. Equipment health diagnosis and prognosis using hidden semi-Markov models. Int. J. Adv. Manuf. Technol. 30 (7–8), 738–749.

Dong, M., Yang, D., Kuang, Y., He, D., Erdal, S., Kenski, D., 2009. PM2. 5 concentration prediction using hidden semi-Markov model-based times series data mining. Expert Syst. Appl. 36, 9046–9055.

Dumont, J., Hernandez, A.I., Fleureau, J., Carrault, G., 2008. Modelling temporal evolution of cardiac electrophysiological features using Hidden Semi-Markov Models. In: Proceedings: Annual International Conference of the IEEE Engineering in Medicine and Biology Society, August 2008, pp. 165–8.

Duong, T.V., Bui, H.H., Phung, D.Q., Venkatesh, S., 2005a. Activity recognition and abnormality detection with the switching hidden semi-Markov model. In: IEEE Computer Society Conference on Computer Vision and Pattern Recognition, 2005 (CVPR 2005), vol. 1, June 20–25, 2005, pp. 838–45.

Duong, T.V., Phung, D.Q., Bui, H.H., Venkatesh, S., 2005b. Efficient coxian duration modelling for activity recognition in smart environments with the hidden semi-Markov model. In: Proceedings of the 2005 International Conference on Intelligent Sensors, Sensor Networks and Information Processing Conference, 2005, December 5–8, 2005, pp. 277–82.

Duong, T.V., Phung, D.Q., Bui, H.H., Venkatesh, S., 2006. Human behavior recognition with generic exponential family duration modeling in the hidden semi-Markov model. In: Proc. of 18th International Conference on Pattern Recognition, 2006 (ICPR 2006), vol. 3, August 20–24, 2006, pp. 202–7.

Economou, T., Bailey, T.C., Kapelan, Z., 2014. MCMC implementation for Bayesian hidden semi-Markov models with illustrative applications. Stat. Comput. 24, 739–752.

Eldar, E., Morris, G., Niv, Y., 2011. The effects of motivation on response rate: a hidden semi-Markov model analysis of behavioral dynamics. J. Neurosci. Methods 201, 251–261.

Elliott, R., Aggoun, L., Moore, J., 1995. Hidden Markov models, estimation and control. Springer-Verlag New York, Inc., New York, NY, USA.

Elliott, R., Limnios, N., Swishchuk, A., 2013. Filtering hidden semi-Markov chains. Stat. Probab. Lett. 83, 2007–2014.

Ephraim, Y., Merhav, N., 2002. Hidden Markov processes. IEEE Trans. Inf. Theory 48 (6), 1518–1569.

Faisan, S., Thoraval, L., Armspach, J.-P., Heitz, F., 2002. Hidden semi-Markov event sequence models: application to brain functional MRI sequence analysis. In: Proceedings of 2002 International Conference on Image Processing, vol. 1, September 22–25, 2002, pp. I-880–I-883.

Faisan, S., Thoraval, L., Armspach, J.-P., Metz-Lutz, M.-N., Heitz, F., 2005. Unsupervised learning and mapping of active brain functional MRI signals based on hidden semi-Markov event sequence models. IEEE Trans. Med. Imaging 24 (2), 263–276.

Ferguson, J.D., 1980. Variable duration models for speech. In: Symp. Application of Hidden Markov Models to Text and Speech, Institute for Defense Analyses, Princeton, NJ, pp. 143–79.

Finesso, L., 1990. Consistent Estimation of the Order for Markov and Hidden Markov Chains (Ph.D. dissertation). Univ. Maryland, College Park.

Ford, J., Krishnamurthy, V., Moore, J.B., 1993. Adaptive estimation of hidden semi-Markov chains with parameterised transition probabilities and exponential decaying states. In: Proc. of Conf. on Intell. Signal Processing and Communication Systems (ISPACS), Sendai, Japan, October 1993, pp. 88–92.

Fox, E.B., 2009. Bayesian Nonparametric Learning of Complex Dynamical Phenomena (Ph.D. thesis). MIT, Cambridge, MA.

Gales, M., Young, S., 1993. The Theory of Segmental Hidden Markov Models. Technical Report CUED/F-INFENG/TR 133. Cambridge University Engineering Department.

Ge, X., Smyth, P., 2000a. Deformable Markov model templates for time-series pattern matching. In: Proc. of the Sixth ACM SIGKDD International Conference on Knowledge Discovery and Data Mining, Boston, MA, 2000, pp. 81–90.

Ge, X., Smyth, P., 2000b. Segmental semi-Markov models for change-point detection with applications to semi-conductor manufacturing. Technical Report UCI-ICS 00-08, <http://www.ics.uci.edu/~datalab/papers/trchange.pdf>, March 2000. <http://citeseer.ist.psu.edu/ge00segmental.html>.

Geramifard, O., Xu, J.-X., Zhou, J.-H., Li, X., 2011. Continuous health condition monitoring: a single hidden semi-Markov model approach. In: 2011 IEEE Conference on Prognostics and Health Management (PHM), June 20–23, 2011, pp. 1–10.

Ghahramani, Z., 2001. An introduction to hidden Markov models and Bayesian networks. Int. J. Pattern Recognit. Artif. Intell. 15 (1), 9–42.

Gillman, M., Kejak, M., Pakos, M., 2014. Learning about rare disasters: implications for consumption and asset prices. February 2014. Working Paper Series, Available online from: <http://www.cerge-ei.cz/publications/working_papers/>.

Glynn, P.W., 1989. A GSMP formalism for discrete event systems. Proc. IEEE 77 (1), 14–23.

Gordon, N.J., Salmond, D.J., Smith, A.F.M., 1993. Novel approach to nonlinear/non-Gaussian Bayesian state estimation. IEE Proc. F Radar Signal Process. 140 (2), 107–113, http://dx.doi.org/10.1049/ip-f-2.1993.0015. Retrieved 2009-09-19.

Gu, H.-Y., Tseng, C.-Y., Lee, L.-S., 1991. Isolated-utterance speech recognition using hidden Markov models with bounded state durations. IEEE Trans. Signal Process., [see also IEEE Trans. Acoustics Speech Signal Process.] 39 (8), 1743–1752.

Guedon, Y., 2003. Estimating Hidden semi-Markov chains from discrete sequences. J. Comput. Graph. Stat. 12 (3), 604–639.

Guedon, Y., 2005. Hidden hybrid Markov semi-Markov chains. Comput. Stat. Data Anal. 49 (3), 663–688.

Guedon, Y., 2007. Exploring the state sequence space for hidden Markov and semi-Markov chains. Comput. Stat. Data Anal. 51 (5), 2379–2409.

Guedon, Y., Cocozza-Thivent, C., 1989. Use of the Derin's algorithm in hidden semi-Markov models for automatic speech recognition. In: Proc. of 1989 International Conference on Acoustics, Speech, and Signal Processing (ICASSP-89), May 23–26, 1989, pp. 282–5.

Guedon, Y., Barthelemy, D., Caraglio, Y., Costes, E., 2001. Pattern analysis in branching and axillary flowering sequences. J. Theor. Biol. 212 (4), 481–520.

Hanazawa, T., Kita, K., Nakamura, S., Kawabata, T., Shikano, K., 1990. ATR HMM-LR continuous speech recognition system. In: Proc. of 1990 International Conference on Acoustics, Speech, and Signal Processing, 1990. ICASSP-90. April 3–6, 1990, pp. 53–6.

He, H., Wu, S., Banerjee, P., Bechhoefer, E., 2006. Probabilistic model based algorithms for prognostics. In: Proc. of 2006 IEEE Aerospace Conference, March 4–11, 2006.

He, J., Leich, H., 1995. A unified way in incorporating segmental feature and segmental model into HMM. In: Proc. of 1995 International Conference on Acoustics, Speech, and Signal Processing, 1995. ICASSP-95, vol. 1, May 9–12, 1995, pp. 532–5.

Henke, D., Smyth, P., Haffke, C., Magnusdottir, G., 2012. Automated analysis of the temporal behavior of the double Intertropical Convergence Zone over the east Pacific. Remote Sens. Environ. 123 (2012), 418–433.

Hieronymus, J.L., McKelvie, D., McInnes, F., 1992. Use of acoustic sentence level and lexical stress in HSMM speech recognition. In Proc. of 1992 IEEE International Conference on Acoustics, Speech, and Signal Processing, 1992. ICASSP-92. vol. 1, March 23–26, 1992, pp. 225–7.

Holmes, W.J., Russell, M.J., 1995. Experimental evaluation of segmental HMMs. In Proc. of International Conference on Acoustics, Speech, and Signal Processing, 1995. ICASSP-95, vol. 1, May 9–12, 1995, pp. 536–9.

Holmes, W.J., Russell, M.J., 1999. Probabilistic-trajectory segmental HMMs. Comput. Speech Lang. 13 (1), 3–37.

Hongeng, S., Nevatia, R., 2003. Large-scale event detection using semi-hidden Markov models. In: Proceedings of Ninth IEEE International Conference on Computer Vision, October 13–16, 2003, pp. 1455–62.

Hongeng, S., Nevatia, R., Bremond, F., 2004. Video-based event recognition: activity representation and probabilistic methods. Comp. Vis. Image Underst. 96, 129–162.

Hu, J., Kashi, R., Wilfong, G., 1999. Document classification using layout analysis. In: Proc. of First Intl. Workshop on Document Analysis and Understanding for Document Databases, Florence, Italy, September 1999.

Hu, J., Kashi, R., Wilfong, G., 2000. Comparison and classification of documents based on layout similarity. Inf. Retr. 2 (2), 227–243.

Huang, X.D., 1992. Phoneme classification using semicontinuous hidden Markov models. IEEE Trans. Signal Process., [see also IEEE Trans. Acoustics Speech Signal Process.] 40 (5), 1062–1067.

Hughes, N.P., Tarassenko, L., Roberts, S.J., 2003. Markov models for automated ECG interval analysis. Adv. Neural Inf. Process. Syst.. Available from: <http://citeseer.ist.psu.edu/hughes03markov.html>.

Hughes, N.P., Roberts, S.J., Tarassenko, L., 2004. Semi-supervised learning of probabilistic models for ECG segmentation. In: Proc. of 26th Annual International Conference of the Engineering in Medicine and Biology Society, 2004. EMBC 2004. vol. 1, 2004, pp. 434–7.

Jiang, R., Kim, M.J., Makis, V., 2012. Maximum likelihood estimation for a hidden semi-Markov model with multivariate observations. Qual. Reliab. Engng. Int. 2012 (28), 783–791.

Johnson, M.T., 2005. Capacity and complexity of HMM duration modeling techniques. IEEE Signal Process. Lett. 12 (5), 407–410.

Johnson, M.J., Willsky, A.S., 2013. Bayesian nonparametric hidden semi-Markov models. J. Mach. Learn. Res. 14 (2013), 673–701.

oo, R., Bertrand, S., Tam, J., Fablet, R., 2013. Hidden Markov models: the best models for forger movements? PLOS ONE 8 (8), e71246.

u, C.-H., Xu, C.-H., 2013. A new method for user dynamic clustering based on hsmm in model f SaaS. Appl. Math. Inf. Sci. 7 (3), 1059–1064.

atagiri, S., Lee, C.-H., 1993. A new hybrid algorithm for speech recognition based on HMM segentation and learning vector quantization. IEEE Trans. Speech Audio Process. 1 (4), 421–430.

e, C.C., Llinas, J., 1999. Literature survey on ground target tracking problems. Research roject Report, Center for Multisource Information Fusion, State University of New York at uffalo.

Khorram, S., Sameti, H., King, S., 2015. Soft context clustering for F0 modeling in HMM-based speech synthesis. EURASIP J. Adv. Signal Process. 2015, 2. Available from: http://asp.eurasipjournals.com/content/2015/1/2>.

Kim, S., Smyth, P., 2006. Segmental hidden Markov models with random effects for waveform modeling. J. Mach. Learn. Res. 7, 945–969.

Kobayashi, H., Yu, S.-Z., 2007. Hidden semi-Markov models and efficient forward–backward algorithms. In: 2007 Hawaii and SITA Joint Conference on Information Theory, Honolulu, Hawaii, May 29–31, 2007, pp. 41–6.

Krishnamurthy, V., Moore, J.B., 1991. Signal processing of semi-Markov models with exponentially decaying states. In: Proceedings of the 30th Conference on Decision and Control, Brighton, England, December 1991, pp. 2744–9.

Krishnamurthy, V., Moore, J.B., Chung, S.H., 1991. Hidden fractal model signal processing. Signal Process. 24 (2), 177–192.

Kullback, S., Leibler, R.A., 1951. On information and sufficiency. Ann. Math. Stat. 22 (1), 79–86, http://dx.doi.org/10.1214/aoms/1177729694.MR 39968.

Kulp, D., Haussler, D., Reese, M.G., Eeckman, F.H., 1996. A generalized hidden Markov model for the recognition of human genes in DNA. In: Proc. 4th Int. Conf. Intell. Syst. Molecular Bio., 1996, pp. 134–42.

Kundu, A., He, Y., Chen, M.-Y., 1997. Efficient utilization of variable duration information in HMM based HWR systems. In: Proceedings of International Conference on Image Processing, 1997, vol. 3, October 26–29, 1997, pp. 304–7.

Kundu, A., He, Y., Chen, M.-Y., 1998. Alternatives to variable duration HMM in handwriting recognition. IEEE Trans. Pattern Anal. Mach. Intell. 20 (11), 1275–1280.

Kwon, O.W., Un, C.K., 1995. Context-dependent word duration modelling for Korean connected digit recognition. Electron. Lett. 31 (19), 1630–1631.

Lanchantin, P., Pieczynski, W., 2004. Unsupervised non stationary image segmentation using triplet Markov chains. In: Proc. of Advanced Concepts for Intelligent Vision Systems (ACVIS 04), Brussels, Belgium, August 31–September 3, 2004.

Lanchantin, P., Pieczynski, W., 2005. Unsupervised restoration of hidden non stationary Markov chain using evidential priors. IEEE Trans. Signal Process. 53 (8), 3091–3098.

Lanchantin, P., Lapuyade-Lahorguea, J., Pieczynski, W., 2008. Unsupervised segmentation of triplet Markov chains hidden with long-memory noise. Signal Process. 88 (5), 1134–1151.

Langrock, R., 2012. Flexible latent-state modelling of old faithful's eruption inter-arrival times in 2009. Aust. N. Z. J. Stat. 54 (3), 261–279.

Langrock, R., Zucchini, W., 2011. Hidden Markov models with arbitrary state dwell-time distributions. Comput. Stat. Data Anal. 55, 715–724.

Langrock, R., King, R., Matthiopoulos, J., Thomas, L., Fortin, D., Morales, J.M., 2012. Flexible and practical modeling of animal telemetry data: hidden Markov models and extensions. Ecology 93 (11), 2336–2342.

Lapuyade-Lahorgue, J., Pieczynski, W., 2006. Unsupervised segmentation of hidden semi-Markov non stationary chains. In: Twenty Sixth International Workshop on Bayesian Inference and Maximum Entropy Methods in Science and Engineering, MaxEnt 2006, Paris, France, July 8–13, 2006.

Lapuyade-Lahorgue, J., Pieczynski, W., 2012. Unsupervised segmentation of hidden semi-Markov non-stationary chains. Signal Process. 92 (1), 29–42.

Lee, C.-H., Soong, F.K., Juang, B.-H., 1988. A segment model based approach to speech recognition. In: Proc. Int'l. Conf. on Acoust., Speech and Signal Processing, pp. 501–4.

Leland, W., Taqqu, M., Willinger, W., Wilson, D., 1994. On the self-similar nature of ethernet traffic (Extended Version). IEEE/ACM Trans. Netw. 2 (1), 1–15.

Levinson, S.E., 1986a. Continuously variable duration hidden Markov models for automatic speech recognition. Comput. Speech Lang. 1 (1), 29–45.

Levinson, S.E., Rabiner, L.R., Sondhi, M.M., 1983. An introduction to the application of the theory of probabilistic functions of a Markov process in automatic speech recognition. B. S. T. J. 62, 1035–1074.

Levinson, S.E., Ljolje, A., Miller, L.G., 1988. Large vocabulary speech recognition using a hidden Markov model for acoustic/phonetic classification. In: Proc. of 1988 International Conference on Acoustics, Speech, and Signal Processing, 1988. ICASSP-88. April 11–14, 1988, pp. 505–8.

Levinson, S.E., Liberman, M.Y., Ljolje, A., Miller, L.G., 1989. Speaker independent phonetic transcription of fluent speech for large vocabulary speech recognition. In: Proc. of 1989 International Conference on Acoustics, Speech, and Signal Processing, ICASSP-89, May 23–26, 1989, pp. 441–4.

Levy, P., 1954. Processes semi-Markoviens. Proc. Int. Congr. Math. (Amsterdam) 3, 416–426.

Li, B.-C., Yu, S.-Z., 2015. A robust scaling approach for implementation of HsMMs. IEEE Signal Process. Lett. 22 (9), 1264–1268.

Li, M., Yu, S.-Z., 2006. A Network-wide traffic anomaly detection method based on HSMM. In: Proc. of 2006 International Conference on Communications, Circuits and Systems Proceedings, vol. 3, June 2006, pp. 1636–40.

Liang, C., Xu, C.-S., Cheng, J., Lu, H.-Q., 2011. TVParser: An automatic TV video parsing method. In: the 24th IEEE Conference on Computer Vision and Pattern Recognition, CVPR 2011, Colorado Springs, CO, USA, June 20–25, 2011, pp. 3377–84.

Lin, D.-T., Liao, Y.-C., 2011. On-line handwritten signature verification using hidden semi-Markov model. Commun. Comput. Inf. Sci. 173 (2011), 584–589.

Lin, H.-P., Tseng, M.-J., Tsai, F.-S., 2002. A non-stationary hidden Markov model for satellite propagation channel modeling. In: Proceedings of 2002 IEEE 56th Vehicular Technology Conference, VTC 2002-Fall, vol. 4, September 24–28, 2002, pp. 2485–8.

Liu, Q.-M., Dong, M., Peng, Y., 2012. A novel method for online health prognosis of equipment based on hidden semi-Markov model using sequential Monte Carlo methods. Mech. Syst. Signal Process. 32 (2012), 331–348.

Liu, X.B., Yang, D.S., Chen, X.O., 2008. New approach to classification of Chinese folk music based on extension of HMM. International Conference on Audio, Language and Image Processing, ICALIP 2008, July 7–9, 2008, pp. 1172–9.

Ljolje, A., Levinson, S.E., 1991. Development of an acoustic-phonetic hidden Markov model for continuous speech recognition. IEEE Trans. Signal Process., [see also IEEE Trans. Acoustics Speech Signal Process.] 39 (1), 29–39.

Lu, W.-Z., Yu, S.-Z., 2006a. An HTTP flooding detection method based on browser behavior. In: Proc. of 2006 International Conference on Computational Intelligence and Security, vol. 2, November 2006, pp. 1151–4.

Maeno, Y., Nose, T., Kobayashi, T., Koriyama, T., Ijima, Y., Nakajima, H., et al., 2014. Prosodic variation enhancement using unsupervised context labeling for HMM-based expressive speech synthesis. Speech Commun. 57, 144–154.

Makino, T., Takaki, S., Hashimoto, K., Nankaku, Y., Tokuda, K., 2013. Separable lattice 2-d HMMs introducing state duration control for recognition of images with various variations. In: 2013 IEEE International Conference on Acoustics, Speech, and Signal Processing, 2013, pp. 3203–7.

Marcheret, E., Savic, M., 1997. Random walk theory applied to language identification. In: Prc of 1997 IEEE International Conference on Acoustics, Speech, and Signal Processing, ICASSP-9 vol. 2, April 21–24, 1997, pp. 1119–22.

Marhasev, E., Hadad, M., Kaminka, G.A., 2006. Non-stationary hidden semi Markov models activity recognition. In: Proceedings of the AAAI Workshop on Modeling Others fro Observations (MOO-06), 2006.

Mark, B.L., Zaidi, Z.R., 2002. Robust mobility tracking for cellular networks. In: Proc. of IEE International Conference on Communications, 2002. ICC 2002, vol. 1, April 28–May 2, 20C pp. 445–9.

McFarland, J.M., Hahn, T.T., Mehta, M.R., 2011. Explicit-duration hidden Markov model inf ence of UP-DOWN states from continuous signals. PLoS One 6 (6), e21606. Available fro http://dx.doi.org/10.1371/journal.pone.0021606, 2011 Jun 28.

McLachlan, G.J., Krishnan, T., 2008. The EM algorithm and extensions, 2nd ed. Wile New York.

Melnyk, I., Yadav, P., Steinbach, M., Srivastava, J., Kumar V., Banerjee, A., 2013. Detection precursors to aviation safety incidents due to human factors. In: 2013 IEEE 13th Internatior Conference on Data Mining Workshops, pp. 407–12.

Meshkova, E., Ansari, J., Riihijarvi, J., Nasreddine, J., Mahonen, P., 2011. Estimating transm ter activity patterns: an empirical study in the indoor environment. In: 2011 IEEE 22 International Symposium on Personal, Indoor and Mobile Radio Communications, pp. 503–8.

Mitchell, C., Jamieson, L., 1993. Modeling duration in a hidden Markov model with the exp nential family. In: Proc. IEEE International Conference on Acoustics, Speech, and Sigr Processing, ICASSP-93, 1993, pp. 331–4.

Mitchell, C., Harper, M., Jamieson, L., 1995. On the complexity of explicit duration HMM IEEE Trans. Speech Audio Process. 3 (2), 213–217.

Mitchell, C.D., Helzerman, R.A., Jamieson, L.H., Harper, M.P., 1993. A parallel implementation a hidden Markov model with duration modeling for speech recognition. In: Proceedings of the Fi IEEE Symposium on Parallel and Distributed Processing, 1993, December 1–4, 1993, pp. 298–30

Moghaddass, R., Zuo, M.J., 2012a. Multi-state degradation analysis for a condition monitor device with unobservable states. In: 2012 International Conference on Quality, Reliability, Ri Maintenance, and Safety Engineering (ICQR2MSE), June 15–18, 2012, pp. 549–54.

Moghaddass, R., Zuo, M.J., 2012b. A parameter estimation method for a condition-monitor device under multi-state deterioration. Reliab. Eng. Syst. Saf. 106, 94–103.

Moghaddass, R., Zuo, M.J., 2014. An integrated framework for online diagnostic and prognos health monitoring using a multistate deterioration process. Reliab. Eng. Syst. Saf. 124, 92–104.

Moore, M.D., Savic, M.I., 2004. Speech reconstruction using a generalized HSMM (GHSM Digit. Signal Process. 14 (1), 37–53.

Murphy, K.P., 2002a. Hidden semi-Markov models (HSMMs). <http://www.ai.mit.e murphyk>, November 2002.

Murphy, K., 2002b. Dynamic Bayesian Networks: Representation, Inference and Learning (P thesis. Dept. Computer Science, UC Berkeley, Computer Science Division.

Nagasaka, S., Taniguchi, T., Hitomi, K., Takenaka, K., Bando, T., 2014. Prediction of n contextual changing point of driving behavior using unsupervised bayesian double articulati analyzer. In: 2014 IEEE Intelligent Vehicles Symposium (IV), June 8–11, 2014. Dearbo Michigan, USA, pp. 924–31.

Nakagawa, S., Hashimoto, Y., 1988. A method for continuous speech segmentation using HM In: Proc. of 9th International Conference on Pattern Recognition, vol. 2, November 14–17, 19 pp. 960–2.

Natarajan, P., Nevatia, R., 2007a. Hierarchical multi-channel hidden semi-Markov models. The Twentieth International Joint Conference on Artificial Intelligence, Hyderabad, India, January 2007, pp. 2562–7.

Natarajan, P., Nevatia, R., 2007b. Coupled hidden semi Markov models for activity recognition. In: IEEE Workshop on Motion and Video Computing, 2007. WMVC '07. February 2007.

Natarajan, P., Nevatia, R., 2008. Online, real-time tracking and recognition of human actions. In: IEEE Workshop on Motion and video Computing, WMVC 2008, January 8–9, 2008, pp. 1–8.

Natarajan, P., Nevatia, R., 2013. Hierarchical multi-channel hidden semi Markov graphical models for activity recognition. Comp. Vis. Image Underst. 117, 1329–1344.

Niwase, N., Yamagishi, J., Kobayashi, T., 2005. Human Walking Motion Synthesis with Desired Pace and Stride Length Based on HSMM. IEICE Trans. Inf. Syst. E88-D (11), 2492–2499.

Nguyen, H.X., Roughan, M., 2013. Rigorous Statistical Analysis of Internet Loss Measurements. IEEE/ACM Trans. Netw. 21 (3), 734–745.

Nose, T., Yamagishi, J., Kobayashi, T., 2006. A style control technique for speech synthesis using multiple regression HSMM. In: Proc. INTERSPEECH 2006-ICSLP, September 2006, pp. 1324–7.

Nose, T., Yamagishi, J., Masuko, T., Kobayashi, T., 2007a. A style control technique for HMM-based expressive speech synthesis. IEICE Trans. Inf. Syst. E90-D (9), 1406–1413.

Nose, T., Kato, Y., Kobayashi, T., 2007b. A speaker adaptation technique for MRHSMM-based style control of synthetic speech. In: Proc. ICASSP 2007, April 2007, vol. IV, pp. 833–6.

Nunn, W.R., Desiderio, A.M., 1977. Semi-Markov processes: an introduction. Cen. Naval Anal.1–30. Available from: <http://cna.org/sites/default/files/research/0203350000.pdf>.

O'Connell, J., Togersen, F.A., Friggens, N.C., Lovendahl, P., Hojsgaard, S., 2011. Combining Cattle Activity and Progesterone Measurements Using Hidden Semi-Markov Models. J. Agric. Biol. Environ. Stat. 16 (1), 1–16.

Odell, J.J., 1995. The use of Context in Large Vocabulary Speech Recognition (Ph.D. dissertation). Cambridge University.

Oliver, G., Sunehag, P., Gedeon, T., 2012. Asynchronous brain computer interface using hidden semi-Markov models. In: 34th Annual International Conference of the IEEE EMBS, San Diego, California USA, August 28–September 1, 2012, pp. 2728–31.

Ostendorf, M., Roukos, S., 1989. A stochastic segment model for phoneme-based continuous speech recognition. IEEE Trans. Acoustics Speech Signal Process. [see also IEEE Trans. Signal Process.] 37 (12), 1857–1869.

Ostendorf, M., Digalakis, V.V., Kimball, O.A., 1996. From HMM's to segment models: a unified view of stochastic modeling for speech recognition.. IEEE Trans. Speech Audio Process. 4 (5), 360–378.

Oura, K., Zen, H., Nankaku, Y., Lee, A., Tokuda, K., 2006. Hidden semi-Markov model based speech recognition system using weighted finite-state transducer. In: IEEE International Conference on Acoustics, Speech, and Signal Processing, 2006. ICASSP 2006. vol. 1, May 14–19, 2006, pp. I-33–I-36.

Park, K., Kim, G.T., Crovella, M.E., 1997. On the effect of traffic self-similarity on network performance. In: Proceedings of SPIE International Conference on Performance and Control of Network Systems, November 1997, pp. 296–310.

Park, S., Chung, W.K., 2014. Decoding surface electromyogram into dynamic state to extract dynamic motor control strategy of human. In: 2014 IEEE/RSJ International Conference on Intelligent Robots and Systems (IROS 2014), September 14–18, 2014, Chicago, IL, USA, pp. 1427–33.

Park, Y.K., Un, C.K., Kwon, O.W., 1996. Modeling acoustic transitions in speech by modified hidden Markov models with state duration and state duration-dependent observation probabilities. IEEE Trans. Speech Audio Process. 4 (5), 389–392.

Park, Y., Yun, S., Yoo, C.D., 2010. Parametric emotional singing voice synthesis. In: 2010 IEEE International Conference on Acoustics Speech and Signal Processing (ICASSP), 2010, pp. 4814–17.

Pavel, M., Hayes, T.L., Adami, A., Jimison, H.B., Kaye, J., 2006. Unobtrusive assessment of mobility. 28th Annual International Conference of the IEEE Engineering In Medicine and Biology Society, New York, NY, August 30–September 3, 2006.

Peng, G., Zhang, B., Wang, W.S.-Y., 2000. Performance of mandarin connected digit recognizer with word duration modeling. ASR2000 – Automatic Speech Recognition: Challenges for the new Millenium, Paris, France, September 18–20, 2000, pp. 140–4.

Peng, Y., Dong, M., 2011. A prognosis method using age-dependent hidden semi-Markov model for equipment health prediction. Mech. Syst. Signal Process. 25 (2011), 237–252.

Phung, D., Duong, T., Bui, H., Venkatesh., S., 2005a. Activitiy recognition and abnormality detection with the switching Hidden Semi-Markov Model. In Int. Conf. on Comp. Vis. & Pat. Recog, 2005.

Phung, D.Q., Duong, T.V., Venkatesh, S., Bui, H.H., 2005b. Topic transition detection using hierarchical hidden Markov and semi-Markov models. In: Proceedings of the 13th Annual ACM international conference, 2005, pp. 11–20.

Pieczynski, W., 2005. Modeling non stationary hidden semi-Markov chains with triplet Markov chains and theory of evidence. In: 2005 IEEE/SP 13th Workshop on Statistical Signal Processing, July 17–20, 2005, pp. 727–32.

Pieczynski, W., 2007. Multisensor triplet Markov chains and theory of evidence. Int. J. Approx. Reason. 45 (1), 1–16.

Pieczynski, W., Desbouvries, F., 2005. On triplet Markov chains. International Symposium on Applied Stochastic Models and Data Analysis, (ASMDA 2005), Brest, France, May 2005.

Pieczynski, W., Hulard, C., Veit, T., 2002. Triplet Markov chains in hidden signal restoration. SPIE's International Symposium on Remote Sensing, Crete, Greece, September 22–27, 2002.

Pikrakis, A., Theodoridis, S., Kamarotos, D., 2006. Classification of musical patterns using variable duration hidden Markov models. IEEE Trans. Audio Speech Lang. Process. [see also IEEE Trans. Speech Audio Process.] 14 (5), 1795–1807.

Pyke, R., 1961a. Markov renewal processes: definitions and preliminary properties. Ann. Math. Stat. 32, 1231–1242.

Pyke, R., 1961b. Markov renewal processes with finitely many states. Ann. Math. Stat. 32, 1243–1259.

Rabiner, L.R., 1989. A tutorial on hidden Markov models and selected application in speech recognition. Proc. IEEE 77 (2), 257–286.

Ramesh, P., Wilpon, J.G., 1992. Modeling state durations in hidden Markov models for automatic speech recognition. In: 1992 IEEE International Conference on Acoustics, Speech, and Signal Processing, ICASSP-92, vol. 1, March 23–26, 1992, pp. 381–4.

Ratnayake, N., Savic, M., Sorensen, J., 1992. Use of semi-Markov models for speaker-independent phoneme recognition. In: 1992 IEEE International Conference on Acoustics, Speech and Signal Processing, 1992. ICASSP-92. vol. 1, March 23–26, 1992, pp. 565–8.

Riska, A., Squillante, M., Yu, S.-Z., Liu, Z., Zhang, L., 2002. Matrix-Analytic Analysis of a MAP/PH/1 Queue Fitted to Web Server Data. In: Latouche, G., Taylor, P. (Eds.), Fourth International Conference on Matrix Analytic Methods in Stochastic Models, in Matrix Analytic Methods: Theory and Applications. World Scientific, Adelaide, Australia, pp. 333–356.

Russell, S., Norvig, P., 2010. Artificial intelligence: a modern approach., 3rd ed. Prentice Hall, ISBN 978-0136042594.

Russell, M.J., 1993. A segmental HMM for speech pattern modelling. In: 1993 IEEE International Conference on Acoustics, Speech, and Signal Processing, ICASSP-93, vol. 2, April 27–30, 1993, pp. 499–502.

Russell, M.J., 2005. Reducing computational load in segmental hidden Markov model decoding for speech recognition. Electron. Lett. 41 (25), 1408–1409.

Russell, M.J., Cook, A., 1987. Experimental evaluation of duration modelling techniques for automatic speech recognition. In: Proc. of IEEE International Conference on Acoustics, Speech, and Signal Processing, 1987, pp. 2376–9.

Russell, M.J., Moore, R.K., 1985. Explicit modelling of state occupancy in hidden Markov models for automatic speech recognition. In: Proc. IEEE Int. Conf. Acoust. Speech Signal Processing, vol. 10, April 1985, pp. 5–8.

Salzenstein, F., Collet, C., Lecam, S., Hatt, M., 2007. Non-stationary fuzzy Markov chain. Pattern Recogn. Lett. 28 (16), 2201–2208.

Sansom, J., Thompson, C.S., 2008. Spatial and temporal variation of rainfall over New Zealand. J. Geophys. Res. 113 (D6).

Sansom, J., Thomson, P., 2001. Fitting hidden semi-Markov models to breakpoint rainfall data. J. Appl. Probab 38A (2001), 142–157.

Sarawagi, S., Cohen, W.W., 2004. Semi-Markov conditional random fields for information extraction. In: Advances in Neural Information Processing Systems 17, NIPS 2004.

Schabus, D., Pucher, M., Hofer, G., 2014. Joint Audiovisual Hidden Semi-Markov Model-Based Speech Synthesis. IEEE J. Sel. Top. Signal Process. 8 (2), 336–347.

Schwarz, G., 1978. Estimating the dimension of a model. Ann. Stat., 461–464.

Schmidler, S.C., Liu, J.S., Brutlag, D.L., 2000. Bayesian Segmentation of Protein Secondary Structure. J. Comp. Biol. 7, 233–248.

Senior, A., Subrahmonia, J., Nathan, K., 1996. Duration modeling results for an on-line handwriting recognizer. In: Proceedings of 1996 IEEE International Conference on Acoustics, Speech, and Signal Processing, 1996. ICASSP-96. vol. 6, May 7–10, 1996, pp. 3482–5.

Sin, B., Kim, J.H., 1995. Nonstationary hidden Markov model. Signal Process. 46, 31–46.

Sitaram, R., Sreenivas, T., 1997. Connected phoneme HMMs with implicit duration modelling for better speech recognition. In: Proceedings of 1997 International Conference on Information, Communications and Signal Processing, ICICS 1997, September 9–12, 1997, pp. 1024–8.

Sitaram, R.N.V., Sreenivas, T.V., 1994. Phoneme recognition in continuous speech using large inhomogeneous hidden Markov models. In: 1994 IEEE International Conference on Acoustics, Speech, and Signal Processing, ICASSP-94, vol. i, April 19–22, 1994, pp. I/41–I/44.

Smith, W.L., 1955. Regenerative stochastic processes. Proc. R. Soc. Ser. A 232 (1955), 6–31.

Squire, K., 2004. HMM-based Semantic Learning for a Mobile Robot (Ph.D. dissertation). University of Illinois at Urbana-Champaign, 2004, Available from: <http://www.ifp.uiuc.edu/~k-squire/thesis/Kevin_thesis_full.pdf>.

Squire, K., Levinson, S.E., 2005. Recursive maximum likelihood estimation for hidden semi-Markov models. In: 2005 IEEE Workshop on Machine Learning for Signal Processing, September 28–30, 2005, pp. 329–34.

u, C., Shen, J.-Y., 2013. A Novel Multi-hidden Semi-Markov Model for Degradation State dentification and Remaining Useful Life Estimation. Qual. Reliab. Engng. Int 2013 (29), 1181–1192.

utton, C., McCallum, A., 2006. An introduction to conditional random fields for relational arning. In: Getoor, L., Taskar, B. (Eds.), Introduction to Statistical Relational Learning. MIT ress. Available from: <http://www.cs.umass.edu/~mccallum/papers/crf-tutorial.pdf>.

Tachibana, M., Yamagishi, J., Masuko, T., Kobayashi, T., 2005. Performance evaluation of style adaptation for hidden semi-Markov model based speech synthesis. In: INTERSPEECH-2005, 2005, pp. 2805–8.

Tachibana, M., Yamagishi, J., Masuko, T., Kobayashi, T., 2006. A style adaptation technique for speech synthesis using hsmm and suprasegmental features. IEICE Trans. Inf. Syst. E89-D(3), 1092–1099.

Tachibana, M., Izawa, S., Nose, T., Kobayashi, T., 2008. Speaker and style adaptation using average voice model for style control in hmm-based speech synthesis. In: Proc. ICASSP 2008, pp. 4633–6.

Tagawa, T., Yairi, T., Takata, N., Yamaguchi, Y., 2011. Data monitoring of spacecraft using mixture probabilistic principal component analysis and hidden semi-Markov models. In: 2011 3rd International Conference on Data Mining and Intelligent Information Technology Applications (ICMiA), October 24–26, 2011, pp. 141–4.

Taghva, A., 2011. Hidden semi-Markov models in the computerized decoding of microelectrode recording data for deep brain stimulator placement. World Neurosurg. 75 (5–6), 758–763.e4.

Takahashi, Y., Tamamori, A., Nankaku, Y., Tokuda, K., 2010. Face recognition based on separable lattice 2-D HMM with state duration modeling. In: 2010 IEEE International Conference on Acoustics, Speech, and Signal Processing, 2010, pp. 2162–5.

Tan, X.R., Xi, H.S., 2008. Hidden semi-Markov model for anomaly detection. Appl. Math. Comput. 205, 562–567.

Taugourdeau, O., Caraglio, Y., Sabatier, S., Guedon, Y., 2015. Characterizing the respective importance of ontogeny and environmental constraints in forest tree development using growth phase duration distributions. Ecol. Model. 300, 61–72.

Ter-Hovhannisyan, V., 2008. Unsupervised and Semi-Supervised Training Methods for Eukaryotic Gene Prediction (Ph.D. dissertation). Georgia Institute of Technology.

Thoraval, L., 2002. Technical Report: Hidden Semi-Markov Event Sequence Models, 2002, Available from: <http://picabia.ustrasbg.fr/lsiit/perso/thoraval.htm>.

Thoraval, L., Carrault, G., Mora, F., 1992. Continuously variable duration hidden Markov models for ECG segmentation. Eng. Med. Biol. Soc. 14, *Proceedings of the Annual International Conference of the IEEE*, vol. 2, pp. 529–530, October 29-November 1, 1992.

Thoraval, L., Carrault, G., Bellanger, J.J., 1994. Heart signal recognition by hidden Markov models: the ECG case. Meth. Inform. Med. 33, 10–14.

Tilmanne, J., Dutoit, T., 2012a. Continuous control of style and style transitions through linear interpolation in hidden Markov model based walk synthesis. In: Gavrilova, M.L., Tan C.J.K. (Eds.), Trans. Comput. Sci. XVI, LNCS 7380, pp. 34–54.

Tilmanne, J., Dutoit, T., 2012b. Walker speed adaptation in gait synthesis. In: Kallmann, M., Bekris, K., (Eds.), MIG 2012, LNCS 7660, pp. 266–77.

Tokdar, S., Xi, P., Kelly, R.C., Kass, R.E., 2010. Detection of bursts in extracellular spike trains using hidden semi-Markov point process models. J. Comput. Neurosci. 29, 203–212.

Tokuda, K., Masuko, T., Miyazaki, N., Kobayashi, T., 2002. Multi-space probability distributio HMM. IEICE Trans. Inf. Syst. 85 (3), 455–464.

Tuan, T., Park, K., 1999. Multiple time scale congestion control for self-similar network traffic Perform. Eval. 36, 359–386.

Tweed, D., Fisher, R., Bins, J., List, T., 2005. Efficient hidden semi-Markov model inference fo structured video sequences. 2nd Joint IEEE International Workshop on Visual Surveillance an Performance Evaluation of Tracking and Surveillance, 2005, October 15–16, 2005, pp. 247–54.

Valentini-Botinhao, C., Yamagishi, J., King, S., Maia, R., 2014. Intelligibility enhancement o HMM-generated speech in additive noise by modifying Mel cepstral coefficients to increase th glimpse proportion. Comput. Speech Lang. 28, 665–686.

Vaseghi, S.V., 1991. Hidden Markov models with duration-dependent state transition probabilities. Electron. Lett. 27 (8), 625–626.

Vaseghi, S.V., 1995. State duration modeling in hidden Markov models. Signal Process. 41 (1), 31–41.

Vaseghi, S.V., Conner, P., 1992. On increasing structural complexity of finite state speech models. In: 1992 IEEE International Conference on Acoustics, Speech, and Signal Processing, 1992, ICASSP-92, vol. 1, March 23–26, 1992, pp. 537–40.

Veeramany, A., Pandey, M.D., 2011. Reliability analysis of nuclear component cooling water system using semi-Markov process model. Nucl. Eng. Des. 241 (5), 1799–1806.

Wang, J.B., Athitsos, V., Sclaroff, S., Betke, M., 2008. Detecting objects of variable shape structure with hidden state shape models. IEEE Trans. Pattern Anal. Mach. Intell. 30 (3), 477–492.

Wang, X., 1994. Durationally constrained training of hmm without explicit state durational pdf. Proceedings of the Institute of Phonetic Sciences. University of Amsterdam, no. 18.

Wang, X., tenBosch, L.F.M., Pols, L.C.W., 1996. Integration of context-dependent durational knowledge into HMM-based speech recognition. In: Proceedings of Fourth International Conference on Spoken Language, 1996. ICSLP 96. vol. 2, October 3–6, 1996, pp. 1073–6.

Wang, X., Wang, H.-W., Qi, C., Sivakumar, A.I., 2014. Reinforcement learning based predictive maintenance for a machine with multiple deteriorating yield levels. J. Comput. Inf. Syst. 10 (1), 9–19.

Wang, Z., Ansari, J., Atanasovski, V., Denkovski, D., et al., 2011. Self-organizing home networking based on cognitive radio technologies. In: 2011 IEEE Symposium on New Frontiers in Dynamic Spectrum Access Networks (DySPAN), May 3–6, 2011, pp. 666–7.

Wellington, C., Courville, A., Stentz, A., 2005. Interacting Markov random fields for simultaneous terrain modeling and obstacle detection. In: Proc. Rob. Sci. Syst.

Wong, Y.F., Drummond, T., Sekercioglu, Y.A., 2014. Real-time load disaggregation algorithm using particle-based distribution truncation with state occupancy model. Electron. Lett. 50 (9), 697–699.

Wu, C.-H., Hsia, C.-C., Liu, T.-H., Wang, J.-F., 2006. Voice conversion using duration-embedded bi-HMMs for expressive speech synthesis. IEEE Trans. Audio Speech Lang. Process., [see also IEEE Trans. Speech Audio Process.] 14 (4), 1109–1116.

Wu, X., Li, L., Rui, X.-M., 2014. Icing load accretion prognosis for power transmission line with modified hidden semi-Markov model. IET Gener. Transm. Distrib. 8 (3), 480–485.

Wu Y., Hong, G.S., Wong, Y.S., 2010. HMM with explicit state duration for prognostics in face milling. In: 2010 IEEE Conference on Robotics Automation and Mechatronics (RAM), June 28–30, 2010, pp. 218–23.

Xi, L.-Q., Fondufe-Mittendorf, Y., Xia, L., Flatow, J., Widom, J., Wang, J.-P., 2010. Predicting nucleosome positioning using a duration hidden Markov model. BMC Bioinformatics 11, 346, <http://www.biomedcentral.com/1471-2105/11/346>.

Xie, B.-L., Zhang, Q.-S., 2012. Application-layer anomaly detection based on application-layer protocols' keywords. In: 2012 2nd International Conference on Computer Science and Network Technology, pp. 2131–5.

Xie, Y., Yu, S.-Z., 2006a. A dynamic anomaly detection model for web user behavior based on HsMM. In: 10th International Conference on Computer Supported Cooperative Work in Design, May 2006, pp. 1–6.

Xie, Y., Yu, S.-Z., 2006b. A novel model for detecting application layer DDoS attacks. In: First International Multi-Symposiums on Computer and Computational Sciences, IMSCCS-06, vol. 2, April 20–24, 2006, pp. 56–63.

Xie, Y., Tang, S.-S., 2012. Online anomaly detection based on web usage mining. In: 2012 IEEE 26th International Parallel and Distributed Processing Symposium Workshops & PhD Forum, pp. 1177–82.

Xie, Y., Tang, S., Tang, C., Huang, X., 2012. An efficient algorithm for parameterizing HsMM with Gaussian and Gamma distributions. Inf. Process. Lett. 112, 732–737.

Xie, Y., Tang, S., Xiang, Y., Hu, J., 2013a. Resisting Web proxy-based HTTP attacks by temporal and spatial locality behavior. IEEE Trans. Parallel Distributed Syst. 24 (7), 1401–1410.

Xie, Y., Tang, S., Huang, X., Tang, C., Liu, X., 2013b. Detecting latent attack behavior from aggregated Web traffic. Comput. Commun. 36, 895–907.

Xu, R.-C., Jiang, T., 2012. Keeping track of position and cell residual dwell time of cellular networks using HSMM structure and Cell-ID information. In: 2012 IEEE International Conference on Communications (ICC), <http://dx.doi.org/10.1109/ICC.2012.6364758>.

Xu, C., Du, C., Zhao, G.F., Yu, S., 2013. A novel model for user clicks identification based on hidden semi-Markov. J. Netw. Comput. Appl. 36, 791–798.

Yamagishi, J., Kobayashi, T., 2005. Adaptive training for hidden semi-Markov model. In: Proceedings of IEEE International Conference on Acoustics, Speech, and Signal Processing, 2005. (ICASSP '05). vol. 1, March 18–23, 2005, pp. 365–8.

Yamagishi, J., Kobayashi, T., 2007. Average-Voice-Based Speech Synthesis Using HSMM-Based Speaker Adaptation and Adaptive Training. IEICE Trans. Inf. Syst. E90-D (2), 533–543.

Yamagishi, J., Ogata, K., Nakano, Y., Isogai, J., Kobayashi, T., 2006. HSMM-based model adaptation algorithms for average-voice-based speech synthesis. In: Proceedings of 2006 IEEE International Conference on Acoustics, Speech and Signal Processing, 2006. ICASSP 2006. vol. 1, May 14–19, 2006, pp. I-77–I-80.

Yamazaki, T., Niwase, N., Yamagishi, J., Kobayashi, T., 2005. Human walking motion synthesis based on multiple regression hidden semi-Markov model. International Conference on Cyberworlds, November 23–25, 2005.

Yang, P., Dumont, G., Ansermino, J.M., 2006. An adaptive cusum test based on a hidden semi-Markov model for change detection in non-invasive mean blood pressure trend. In: Proceedings of the 28th IEEE EMBS Annual International Conference, New York City, USA, August 30–September 3, 2006, pp. 3395–8.

Yoma, N.B., McInnes, F.R., Jack, M.A., 1998. Weighted Viterbi algorithm and state duration modelling for speech recognition in noise. In: Proceedings of the 1998 IEEE International Conference on Acoustics, Speech, and Signal Processing, 1998. ICASSP '98. vol. 2, May 12–15, 1998, pp. 709–12.

Yoma, N.B., McInnes, F.R., Jack, M.A., Stump, S.D., Ling, L.L., 2001. On including temporal constraints in viterbi alignment for speech recognition in noise. IEEE Trans. Speech Audio Process. 9 (2), 179–182.

Yoma, N.B., Sanchez, J.S., 2002. MAP speaker adaptation of state duration distributions for speech recognition. IEEE Trans. Speech Audio Process. 10 (7), 443–450.

Yu, S.-Z., 2005. Multiple tracking based anomaly detection of mobile nodes. In: 2nd International Conference on Mobile Technology, Applications and Systems, 2005, November 15–17, 2005, pp. 5–9.

Yu, S.-Z., Kobayashi, H., 2003a. An efficient forward-backward algorithm for an explicit duration hidden Markov model. IEEE Signal Process. Lett. 10 (1), 11–14.

Yu, S.-Z., Kobayashi, H., 2003b. A Hidden semi-Markov model with missing data and multiple observation sequences for mobility tracking. Signal Process. 83 (2), 235–250.

Yu, S.-Z., Kobayashi, H., 2006. Practical Implementation of an Efficient Forward-Backward Algorithm for an Explicit-Duration Hidden Markov Model. IEEE Trans. Signal Process. 54 (5), 1947–1951.

Yu, S.-Z., Liu, Z., Squillante, M., Xia, C., Zhang, L., 2002. A hidden semi-Markov model for web workload self-similarity. In: 21st IEEE International Performance, Computing, and Communications Conference (IPCCC 2002), Phoenix, Arizona, April 3–5, 2002, pp. 65–72.

Yu, S.-Z., Mark, B.L., Kobayashi, H., 2000. Mobility tracking and traffic characterization for efficient wireless internet access, IEEE MMT'2000, Multiaccess, Mobility and Teletraffic in Wireless Communications, vol. 5. Duck Key, Florida.

Yun, Y.-S., Oh, Y.-H., 2000. A segmental-feature HMM for speech pattern modeling. IEEE Signal Process. Lett. 7 (6), 135–137.

Yurur, O., Liu, C., Perera, C., Chen, M., Liu, X., Moreno, W., 2014. Energy-efficient and context-aware smartphone sensor employment. IEEE Trans. Vehicular Technol. Available from: http://dx.doi.org/10.1109/TVT.2014.2364619.

Zappi, V., Pistillo, A., Calinon, S., Brogni, A., Caldwell, D., 2012. Music expression with a robot manipulator used as a bidirectional tangible interface. EURASIP J. Audio Speech Music Process. 2, <http://asmp.eurasipjournals.com/content/2012/1/2>.

Zen, H., Tokuda, K., Masuko, T., Kobayashi, T., Kitamura, T., 2004. Hidden semi-Markov model based speech synthesis. In: Proc. of 8th International Conference on Spoken Language Processing, ICSLP, Jeju Island, Korea, October 4–8, 2004, pp. 1393–6.

Zen, H., Tokuda, K., Masuko, T., Kobayashi, T., Kitamura, T., 2007. A hidden semi-Markov model-based speech synthesis system. IEICE Trans. Inf. Syst. E90-D (5), 825–834.

Zha, Y.-B., Yue, S.-G., Yin, Q.-J., Liu, X.-C., 2013. Activity recognition using logical hidden semi-Markov models. In: 2013 10th International Computer Conference on Wavelet Active Media Technology and Information Processing (ICCWAMTIP), 17–19 December 2013, pp. 77–84.

Zhang, W., Chen, F., Xu, W., Zhang, E., 2006. Real-time video intelligent surveillance system. In: 2006 IEEE International Conference on Multimedia and Expo, July 2006, pp. 1021–4.

Zhang, W., Chen, F., Xu, W., Du, Y., 2008. Learning human activity containing sparse irrelevant events in long sequence. In: 2008 Congress on Image and Signal Processing, CISP'08, 2008, pp. 211–15.

Zhao, F., Wang, G.-N., Deng, C.-Y., Zhao, Y., 2014. A real-time intelligent abnormity diagnosis platform in electric power system. In: 2014 16th International Conference on Advanced Communication Technology, (ICACT), February 16–19, 2014, pp. 83–7.

Zhao, Y., Liu, X., Gan S., Zheng, W., 2010. Predicting disk failures with HMM- and HSMM-based Approaches. In: Proc. Industrial Conf. on Data Mining '10, 2010.

Printed in the United States
By Bookmasters